INTRODUCTION TO POWER ANALYSIS

Two-Group Studies

INTRODUCTION TO POWER ANALYSIS
Two-Group Studies

E. C. Hedberg
NORC at the University of Chicago

Los Angeles | London | New Delhi
Singapore | Washington DC | Melbourne

Quantitative Applications in the Social Sciences

A SAGE PUBLICATIONS SERIES

Quantitative Applications in the Social Sciences

A SAGE PUBLICATIONS SERIES

Los Angeles | London | New Delhi
Singapore | Washington DC | Melbourne

FOR INFORMATION:

SAGE Publications, Inc.
2455 Teller Road
Thousand Oaks, California 91320
E-mail: order@sagepub.com

SAGE Publications Ltd.
1 Oliver's Yard
55 City Road
London EC1Y 1SP
United Kingdom

SAGE Publications India Pvt. Ltd.
B 1/I 1 Mohan Cooperative Industrial Area
Mathura Road, New Delhi 110 044
India

SAGE Publications Asia-Pacific Pte. Ltd.
33 Pekin Street # 02-01
Far East Square
Singapore 048763

Acquisitions Editor: Helen Salmon
Content Development Editor: Chelsea Neve
Editorial Assistant: Megan O'Heffernan
Production Editor: Kelly DeRosa
Copy Editor: QuADS Prepress (P) Ltd
Typesetter: QuADS Prepress (P) Ltd
Proofreader: Scott Oney
Indexer: Jeanne Busemeyer
Cover Designer: Candice Harman
Marketing Manager: Susannah Goldes

Printed in the United States of America

Library of Congress Cataloging-in-Publication Data

Names: Hedberg, E. C. (Eric Christopher), 1978- author.

Title: Introduction to power analysis : two-group studies / E. C. Hedberg, NORC at the University of Chicago.

Description: Thousand Oaks, California : Sage Publications, Ltd., 2017. | Series: Quantitative applications in the social sciences ; 176 | Includes bibliographical references.

Identifiers: LCCN 2017037101 | ISBN 9781506343129 (pbk.)

Subjects: LCSH: Statistical power analysis. | Mathematical statistics. | Multivariate analysis.

Classification: LCC QA277 .H43 2017 | DDC 519.5/6–dc23 LC record available at https://lccn.loc.gov/2017037101

MIX
Paper from responsible sources
FSC® C014174

This book is printed on acid-free paper.

17 18 19 20 21 10 9 8 7 6 5 4 3 2 1

CONTENTS

SERIES EDITOR'S INTRODUCTION

I am very pleased to introduce E. C. Hedberg's volume *Introduction to Power Analysis: Two-Group Studies*. Statistical power is the probability of a statistically significant test, assuming the null hypothesis is false and the alternative hypothesis is true. It is a critical component of study design, one that is reviewed and evaluated by institutional review boards (IRBs), funding agencies, and publication outlets. IRBs want to know whether the study is sufficiently powered to justify risk to participants. Funding agencies share a concern that study design, including statistical power, is adequate to meet scientific objectives. In addition, given data collection costs, they are also concerned about cost-effectiveness, specifically that samples not be larger than they need to be. Finally, statistical power is of concern to journals and other academic publication outlets because it is central to robustness and the reliability and reproducibility of results.

This volume introduces statistical power through two-group comparisons, a treatment group and a control group. It is designed to be a supplement for a graduate class in quantitative methods. It also serves as a valuable resource for students, faculty, and other practitioners interested in learning about statistical power. The volume is self-contained, providing all the information needed, including a review of the main statistical distributions, hypothesis testing, and types of errors. The volume is well organized and tightly structured.

Hedberg's strategy is to lay the foundation by conveying the important points through detailed examples for the simplest cases, extending to more complex cases but not in the same detail, preparing readers to think about their own applications. After introducing the what, why, and when of power analysis (Chapter 1) and providing a quick review of key distributions (Chapter 2), he tackles topics in hypothesis testing and power analysis when the population standard deviation is known (Chapter 3) and when it must be estimated (Chapter 4). He then turns to incorporating covariates when testing for differences in group means in balanced designs (Chapter 5), two-level cluster randomized trials (Chapter 6), and two-level multisite randomized trials (Chapter 7). Example power analyses using the major software packages (SPSS, Stata, and R), complete with code and output, can be found online at **study.sagepub.com/hedberg**.

The next two chapters speak to the practicalities of power analysis in ways that are quite helpful to the practitioner. Chapter 8 reviews the assumptions needed to undertake power analysis. Hedberg sums up the chapter this way: "While researchers can never predict the parameters of future data with certainty, . . . it is never a good idea to guess about these assumptions, and using conventions counts as guessing." Hedberg offers specific guidance on how to use previous studies for power analysis assumptions, including studies that do not match exactly the variable to be used. Chapter 9 provides suggestions about how to write about power in a succinct but complete way, a skill needed when explaining study design to IRBs, funding agencies, and peer-reviewed journals. The volume concludes with a brief introduction to more advanced topics, pointing to literature where interested readers can learn more.

Nowadays, there is increasing interest in the use of large datasets consisting of administrative records, transactions, online activity and social media interactions, and GPS tracking for social and behavioral science. When datasets are very large, statistical power is less of an issue (although there may be other aspects of study design that challenge the validity of results). However, even in the age of "big data," there will still be a need for small N studies, and for these, the lessons contained in this volume are important to master. I have enjoyed reading this volume, and I hope you do too.

—*Barbara Entwisle*
Series Editor

PREFACE

When a field of study relies on statistical inference as its evidence base, the reliability of the results, both positive and not, depends a great deal on statistical power. When statistical power is typically low, knowledge can be lost and random noise can become knowledge. This, of course, is an oversimplification to make a simple point: Power is important.

This book is an introduction to power analysis. I do not take this title lightly, as I truly mean for it to be an introduction to the core elements of power analysis. I do this through a case study of comparing the means of two groups, and walking the reader through the simple and complex considerations of this research question. From covariates to clusters, comparing two groups can sometimes be a complicated task, and so can the power analysis.

Through this introduction, I hope readers will gain insight into the analyses they perform, and through that insight, gain a foundation to understand the power analyses that apply to their research.

As for the small details, I have tried to keep the notation consistent. Statistics is an abstract language, with a dialect for every teacher. Thus, while I cannot keep my notation for, say, multilevel modeling consistent with both Raudenbush and Hedges at the same time, I can try to at least keep it consistent between Hedberg and Hedberg. This consistency does mean that my notation will be different than others' notation. I have tried to footnote major deviations from convention. In short, be aware that, for example, what I mean by τ is not what Raudenbush means by τ.

In this volume, I use a lot of examples. Some of these examples are analyses that involve real data. Please understand that the data used in the examples are for pedagogical purposes. While the data are real, they have been sampled from the larger datasets to make points about power analysis. Thus, the results should not be cited as empirical evidence. I am indebted to the researchers who put their data on the online repositories for writers like me to sift through at night.

Finally, to adequately cite intellectual contributions, this book was made far easier to write and lay out in LaTeX with the "texreg" package for R (Leifeld, 2013), the "xtable" package for R (Dahl, 2009), and the R package "tikzDevice" (Sharpsteen & Bracken, 2013) for graphics.

I hope you enjoy this volume and learn a little bit about statistical power. Example power analyses using the major software packages (SPSS, Stata, and R), complete with code and output, can be found online at **study.sagepub.com/hedberg**.

ABOUT THE AUTHOR

E. C. Hedberg is a Senior Research Scientist at NORC at the University of Chicago. He received his undergraduate degree in sociology from the University of Minnesota and his PhD in sociology from the University of Chicago. His work is primarily focused on estimating design parameters useful for power analysis, multilevel modeling, social capital theory, and evaluation research.

ACKNOWLEDGMENTS

The research reported here was supported by the Institute of Education Sciences, U.S. Department of Education, through Grant R305D140019, NORC at the University of Chicago. The opinions expressed are those of the authors and do not necessarily represent the views of the Institute or the U.S. Department of Education.

The author would like to thank the following individuals for their support and comments while writing this manuscript:

Larry Hedges, Northwestern University
Charles Katz, Arizona State University
Arend Kuyper, Northwestern University
Danielle Wallace, Arizona State University

The author would also like to acknowledge the contributions of the following reviewers:

Chris Aberson, Humboldt State University
Leslie Echols, Missouri State University
Erin M. Fekete, University of Indianapolis
Stephanie J. Jones, Texas Tech University
Karin Lindstrom Bremer, Minnesota State University, Mankato
Fred Oswald, Rice University
Jason Popan, University of Texas–Pan American
Gary Popoli, Stevenson University
Ben Kelcey, University of Cincinnati
Bryan J. Rooney, Concordia University of Edmonton

CHAPTER 1. THE WHAT, WHY, AND WHEN OF POWER ANALYSIS

What Is Statistical Power?

Statistical power is the probability of a statistically significant test (assuming the null hypothesis is in fact false and the alternative hypothesis is true), given several factors that include the acceptable level of uncertainty, the size of the effect, the sample design, and sample size (Cohen, 1988). Power levels near zero indicate that there is little chance of detecting an effect, whereas power levels near one indicate a high probability of detecting an effect. The convention in the social sciences is to design studies with a power of at least 0.8, or an 80% chance of detecting an effect. Power is directly related to the types of errors in statistical tests and the viability of the knowledge base in a research field.

Errors in Research

For any alternative hypothesis (e.g., exercise reduces weight), relative to a null hypothesis (e.g., exercise does not affect weight), there are four possible outcomes. The first outcome is that the researcher concludes that the alternative hypothesis is likely when in fact it is true. These are generally termed "significant" results. For example, the researcher concludes that a new curriculum is effective in increasing math scores when in fact the curriculum does increase scores.

The second outcome is when the researcher concludes that the null hypothesis is likely true when in fact the null hypothesis is true. These are generally called "null" results. For example, the researcher concludes that a new drug has no impact on headaches when in fact the drug does not have an impact on headaches. These two outcomes, "significant" and "null," are ideal because the researcher's inference corresponds with reality.

The next two outcomes of research are errors. The first type of errors are Type I errors, which occur when the researcher rejects the null hypothesis in favor of the alternative hypothesis when in fact the null is true. For example, a Type I error is to conclude that a therapy has an influence on criminal behavior when in fact the therapy does not influence behavior. The convention in research is to accept that this happens 5% of the time, which corresponds to a Type I error of 0.05. We use the Greek letter alpha (α) as the Type I error symbol and note that our test corresponds to $\alpha = 0.05$.

The second type of errors are Type II errors, which occur when the researcher accepts the null hypothesis when in fact the alternative is true. For example, a Type II error is to conclude that a policy has no impact on the economic outcomes of the poor when in fact the policy does affect economic outcomes. The convention is to hope studies have no more than a 20% chance of this happening, which corresponds to a Type II error of 0.2. We use the Greek letter beta (β) as the Type II error symbol and note that our test corresponds to $\beta = 0.2$. The power of a test is $1 - \beta$, or the chance that a test detects an effect when in fact it is true.

A Small Simulation

In some cases, researchers use Monte Carlo simulations to determine power. Simulations can sound daunting, but the premise is a simple one. Using random number generators available in most statistical software packages, a researcher enters into a program various assumptions. For example, a researcher may assume that variables are distributed normally with a mean of 0 and a standard deviation of 1, both of which are entered into the random number generator to produce a set of cases (also under the control of the researcher). Once the random numbers (i.e., random observations) are produced, the researcher simply performs a typical analysis on the simulated data (the software does not mind that the data are fake) and records the resulting test statistic and other parameters of interest. This is done several thousand times, with each simulation producing a slightly different result.

The set of results from the simulations can be plotted as a histogram to reveal the simulated sampling distribution (which is a nice way to test the central limit theorem). In terms of power analysis, the researcher can also tally the number of simulations in which the result was statistically significant. The proportion of significant results is the power of the set of assumptions (i.e., the test) used to drive the simulations.

For example, we can illustrate the types of errors with the following simulations of hypothesis tests (which will be reviewed throughly throughout this volume). The first simulation draws 2 sets of 50 cases, each from a normal distribution with a mean of 0 and a standard deviation of 1. Because both sets come from the same distribution (standard normal with a mean of 0 and standard deviation of 1), we know that the true difference in group means is 0. Once the groups are created, a simple two-tailed t test is performed to determine whether the groups have

Figure 1.1 Histogram of 10,000 simulated two-tailed tests ($\alpha = 0.05$) with a sample of 100 cases where the true difference between groups is 0 standard deviations (i.e., null hypothesis is true); incorrect results (i.e., Type I error) are shaded black.

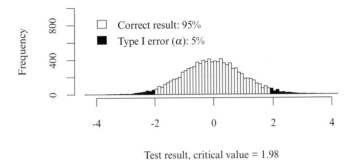

Test result, critical value = 1.98

the same mean. We play this game 10,000 times and record the resulting t test in a dataset. Figure 1.1 presents a histogram of these simulations. The critical values for such a test when we set $\alpha = 0.05$ are about -1.98 and 1.98. Figure 1.1 shows that about 5% of the time the test result is in the critical region, which corresponds to a Type I error rate of $\alpha = 0.05$.

Next, consider another simulation. In this simulation, we draw one group from a distribution with a mean of 0 and a standard deviation of 1, and then draw another group from a distribution with a mean of 0.5 and a standard deviation of 1. Thus, we know that the true difference in means is 0.5 standard deviations. We play this game 10,000 times and record the t test as before. Figure 1.2 presents a histogram of these simulations. Again, the critical values for such a test when we set $\alpha = 0.05$ are about -1.98 and 1.98. Figure 1.2 shows that about 30% of the time, the test result is between the critical regions, which corresponds to a Type II error rate of $\beta = 0.3$, or power of about 0.7. If we were to change the parameters of the simulation, such as the sample size or effect size, we would get a different result from the simulations.

This volume is about using assumptions and formulas to make a priori determinations about the power of a test before the data are collected and without simulations.

Why Should Power Be a Consideration When Planning Studies?

There are two reasons why understanding the power of a test before the data are collected is important. The first reason power analyses are

4

Figure 1.2   Histogram of 10,000 simulated two-tailed tests with a sample of 100 cases where the true difference between groups is 0.5 standard deviations (i.e., alternative hypothesis is true); incorrect results (i.e., Type II error) are shaded gray.

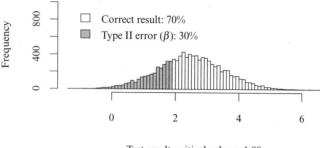

Test result, critical value = 1.98

important is simple: Rational researchers should wish to maximize the chance of detecting the hypothesized effect while minimizing the potential risks to participants. This reason originates from both economic and ethical considerations. The economic consideration is that studies cost money, and studies in which the outcome is inconclusive are a waste of resources. The ethical consideration is that if the intervention poses a risk to participants, the research should involve only the minimum necessary sample in order to achieve the desired power.

Therefore, an a priori estimate of the chance to detect an effect going into an experiment provides useful information about the probable success and risk of the project. If a power analysis indicates that the chance of success is low, it provides an opportunity for the research team to reevaluate their plans and change the design, the sample, or both. If power is very high and the intervention risky, the power analysis can provide some justification for reducing the treatment sample size. Also, researchers who are hypothesizing to *not* find an effect must always be aware of power. They should adequately power the study for the smallest meaningful effect. In doing so, if they do not find a statistically significant difference, they are confident that the result is based on the actual lack of an effect and not simply an underpowered study.

The second reason is a little more complex, but equally important. Recently, several publications have outlined the disturbing notions that many of the published findings in the scientific literature are not true (e.g., Ioannidis, 2005) and that many findings are not replicated (e.g.,

Open Science Collaboration, 2015). There are many reasons for this, and most are beyond the scope of this volume. One reason, however, is low statistical power among many studies.

Consider again the four outcomes of any test: significant, null, Type I error, and Type II error. Using formulas worked out by Ioannidis (2005), and making some assumptions, we can estimate the proportion of research results that fall into each of the four types of outcomes. In the following formulas, R is the ratio of true hypotheses tested to the false hypotheses tested, α is the Type I error rate, and β is the Type II error rate. In the simplest table from Ioannidis's (2005) article, the expected percentage of "significant" findings is

$$\text{Significant} = 100 \times \frac{(1-\beta)R}{R+1}. \qquad \text{(Equation 1.1)}$$

The percentage of "null" findings is

$$\text{Null} = 100 \times \frac{1-\alpha}{R+1}. \qquad \text{(Equation 1.2)}$$

The percentage of findings that are Type I errors is

$$\text{Type I errors} = 100 \times \frac{\alpha}{R+1}. \qquad \text{(Equation 1.3)}$$

Finally, the percentage of findings that are Type II errors is

$$\text{Type II errors} = 100 \times \frac{\beta R}{R+1}. \qquad \text{(Equation 1.4)}$$

Suppose that of all hypotheses tested, 20% are actually true. This means that the ratio of true to false hypotheses is $R = 0.2/0.8 = 0.25$. Next, suppose that all hypotheses are tested with $\alpha = 0.05$. This means that the percentage of null hypotheses is about $100 \times \frac{1-0.05}{0.25+1} = 76\%$. This also means that the number of Type I errors is about $100 \times \frac{0.05}{0.25+1} = 4\%$. Thus, the total percentage of false hypotheses is 80% (or, conversely, the number of true hypotheses is 20% as previously stated). Figure 1.3 reports this 80% with the Null and Significant slices of the pie chart.

To illustrate the implications of power, first assume that research is published only if the findings appear to be significant (to support the idea that significant findings are more likely to be published see, e.g., Easterbrook, Gopalan, Berlin, & Matthews, 1991). Next, suppose that all research was conducted with power 0.8 ($\beta = 0.2$, see the left chart in Figure 1.3). This means that out of all hypotheses, $100 \times \frac{0.2 \times 0.25}{0.25+1} = 4\%$ are

Figure 1.3   Percentage of hypothetical hypotheses that are significant ("Sig"),
Type I errors ("α"), null ("Null"), and Type II errors ("β") by
power levels of 0.8 and 0.5. This assumes that 20% of tested
hypotheses are actually true and are tested with $\alpha = 0.05$.

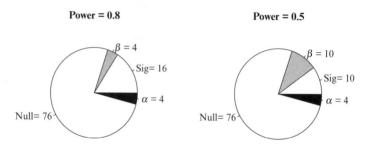

Type II errors, and $100 \times \frac{(1-0.2) \times 0.25}{0.25 + 1} = 16\%$ of all hypotheses tested are
significant. This means a fifth of true hypotheses tested are *not published*,
while another fifth of *published* hypotheses are actually false.

The situation is worse when tests are conducted with lower power, for
example, when power is 0.5 ($\beta = 0.5$, see right side of Figure 1.3). The null
hypotheses are not affected, but now the percentage of all hypotheses that
are Type II errors is 10%, and the percentage of all hypotheses that are
truly significant findings is also 10%. This means that 14% of hypotheses
are published, and almost 30% of those published findings are actually
false compared with 20% in the scenario with higher power.

The point of this exercise was to show that as power drops in a field of
study, the proportion of published results that are due to chance, which
are actually null, increases. It has been demonstrated that in some fields,
power has been low at times (see, e.g., Spybrook, 2007). This is one reason that some researchers are skeptical about the body of knowledge currently produced. One way to increase the confidence in a body of knowledge is to increase the power of the studies from which that body is built.

When Should You Perform a Power Analysis?

Succinctly, power analyses are most useful before data collection. Once
the data are collected, a power analysis is little more than a postmortem
for insignificant results (if the results are significant, then power was
obviously sufficient). However, useful insights can still be gained from
power analyses for observational data.

Experiments

Power analysis is most useful (and easier) for randomized experiments. As we will see in the later chapters, correlations between the variable of interest (i.e., treatment group membership) and other covariates complicates a priori expectations about statistical tests. This is because information about the interrelationships among all covariates must be taken into account. When the variable of interest is randomized, these correlations (in expectation) drop out leaving relatively simple formulas for power analyses.

Observational Studies

In observational studies, it is not power analysis but precision analysis that is usually an activity before data collection. Precision analysis is similar to power analysis, but instead of estimating the likelihood of detecting a significant effect, like the difference in means between two groups, precision analysis determines the precision (e.g., margin of error) for an estimate of the population mean, total, or percentage given a sample size and design (e.g., see the discussion of determining sample size in Lohr, 2009). Also, as in the different power analyses discussed throughout this volume, researchers can also find the necessary sample to achieve a level of precision given a confidence level. Such analyses are critical in planning surveys with univariate analysis goals, such as opinion polls.

Again, power (and precision) analysis is not very useful once the data have been collected. If a result is statistically significant, a power analysis showing low power does not negate the results. If the result is not significant, it is sometimes useful to determine, given the achieved effect size, how large a sample would need to be in order to gain a significant result. If this hypothetical sample size is close to the one observed, then a plausible argument could be made that the study was unlucky and such an effect may exist. On the other hand, if the sample would have to be much larger than is plausible to collect, than an argument could be made that a meaningful effect simply does not exist.

Significance and Effect

Power is about statistical significance. However, key elements of power analysis involve the interplay between effect size (which can be thought of as the practical significance) and sample size. For years, many scientific fields were too narrowly focused on the ability to say that some effect was likely nonzero and the expense of looking at the size of the effect (see, e.g., Ziliak & McCloskey, 2008). These are different quantities. As you will discover through reading this volume, a large enough sample will be able to detect even the smallest effect with precision (i.e., with statistical significance). However, this may mean that the research accomplished a hollow achievement: the precise estimate of nothing important. The reader is encouraged to keep this in mind when performing power analyses, as it should always be the goal of research to detect (or not) a meaningful effect that has practical significance.

What Do You Need to Know to Perform a Power Analysis?

Power analyses for different tests require different elements. In this volume, the focus is on the difference between two group means using linear regression. This case study will focus on understanding the components of the test, which parameters make up those components, what can be assumed, and what is under the researcher's control. In general, any power analysis for a parametric test needs to understand the following:

- How the population parameter (e.g., the difference in group means) is estimated

- How the sampling variance of the population parameter is calculated (the square root of which is the standard error)

- How the test is calculated

- Which parts of the test can be converted into scale-free parameters

- How to calculate areas under the sampling distribution curves of the null and alternative hypotheses (typically with a computer)

- How to rearrange the test statistic to isolate other factors that influence power, such as the effect size or sample size

This volume is a guide to how to understand these six things in the context of looking at the difference between group means. The hope is that in working through these examples readers can explore how to perform and understand a power analysis for their own work.

The Structure of the Volume

The next chapter will be a brief review of how the main statistical distributions used in hypothesis testing (the standard normal, t, χ^2, and F) relate to the standard normal distribution and to each other through the normal distribution.

Following that is a chapter focusing on basic topics in hypothesis testing and types of errors, and how they relate to simple power analyses for testing the difference between two group means. The bulk of the volume then focuses on comparing two groups in simple random samples, when using covariates, and in more complex samples.

The volume then ends with a discussion on how to gather the assumptions for power analysis and how to write about your power analysis. The concluding chapter offers some next steps and examples of more advanced texts on power.

Summary

In this introductory chapter, we explored a general definition of power. Through a simulation, we approximated a sampling distribution of tests with known outcomes. This illustrated the Type I and Type II errors of the test of the difference between group means.

Next, we reviewed reasons why power analysis is important, focusing on how power analysis is essential for any discipline to be assured that most of the published results are in fact true, which does not appear to be the state of affairs in many areas of inquiry.

Finally, this chapter concluded with a look ahead to the rest of the volume.

CHAPTER 2. STATISTICAL DISTRIBUTIONS

In this brief chapter, the basics of the major statistical distributions are discussed. Instead of a focus on the derivation of individual distributions, the goal of this chapter is to present the reader with the *relationships* of the major distributions with the standard normal distribution.

Understanding these relationships is important to recognize the key distributions in this volume: the t and F distributions. In the later chapters, we work with more complex samples that force us to work with two-way analysis of variance (ANOVA) models (where one factor is a sample of groups). Knowing the relationships from this chapter allows us to simplify the problem of complex sample F tests into t tests.

Normally Distributed Random Variables

We start with a discussion of the normal distribution. The *central limit theorem* (Rice, 2006) states that many statistical sampling distributions are normally distributed, or are distributed in such a way that they are related to the normal distribution. This is why the "normal" curve is always emphasized in introductory courses. As you saw in the simulations in Chapter 1, the test statistic histograms had a "bell" shape to them (e.g., Figure 1.1).

While many variables in datasets have roughly normally distributed variables, such as height or weight, the focus here is the distribution of sample statistics. That is, the variable distribution is not a column in a dataset, but is instead a hypothetical distribution of a resulting statistic from many samples. This distribution of samples' results is often called a "sampling distribution."

The difference between group means for some variable Y, as observed in a collection of data about each group, is a random variable itself because the sample is random (Lohr, 2009). If we collected another random sample, the observed difference would be a different result. Another sample, another different result, and so on. This makes our statistic of interest, the difference between group means, a random variable.

Like any other continuous random variable, we can express a density function to describe the shape of the distribution. For the normal distri-

Figure 2.1   The standard normal distribution where $\mu = 0$ and $\sigma = 1$.

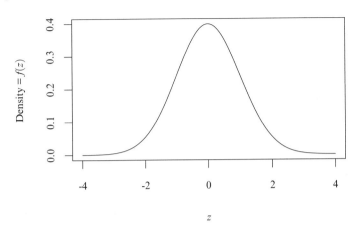

bution, this density function $f(z)$ is (Rice, 2006)

$$f(z) = \frac{1}{\sigma\sqrt{2\pi}}e^{-(z-\mu)^2/(2\sigma^2)}. \qquad \text{(Equation 2.1)}$$

The key parameters of this function are the mean of the distribution (μ) and its standard deviation σ. If a variable X is distributed normally, we note this as $X \sim N(\mu, \sigma)$. Thus, for every point along the number line z, we can use Equation 2.1 to find the density value. Figure 2.1 plots Equation 2.1 where $\mu = 0$ and $\sigma = 1$, also called the standard normal ($Z \sim N(0,1)$).

Density functions have the attractive property in that the total area under the curve is 1. This allows us to portion the distribution in such a way as to say, "the proportion of values before z is . . ." or "the proportion of values after z is . . ." or "the proportion of values between z_a and z_b is . . ." We can do this using the cumulative distribution function (CDF) for the normal distribution (typically with tables or a computer), which we note as Φ.

For example, if we wanted to find the probability that the population of X is less than or equal to x, and we know that the population mean is μ and standard deviation is σ, we can express this using the CDF as

$$P(X \leq x) = \Phi\left(\frac{x-\mu}{\sigma}\right). \qquad \text{(Equation 2.2)}$$

If we wanted to find the probability that the population of X is greater than or equal to x, we can express this using the CDF as

$$P(X \geq x) = 1 - \Phi\left(\frac{x - \mu}{\sigma}\right). \qquad \text{(Equation 2.3)}$$

If we wanted to find the probability that the population falls between x_a and x_b, assuming that $x_a < x_b$, we can express this using the CDF as well:

$$P(x_a \leq X \leq x_b) = \Phi\left(\frac{x_b - \mu}{\sigma}\right) - \Phi\left(\frac{x_a - \mu}{\sigma}\right). \qquad \text{(Equation 2.4)}$$

Of course, $\frac{x-\mu}{\sigma}$ is the familiar formula for the z-score. Most statistical packages have functions for the cumulative normal distribution, and power analysis makes good use of them. For example, $\Phi(1.96) = 0.975$, $1 - \Phi(1.96) = 0.025$, $\Phi(1.96) - \Phi(-1.96) = 0.950$, and $1 - (\Phi(1.96) - \Phi(-1.96)) = 0.05$ (see Figure 2.2).

Normally distributed sampling distributions work much the same way. They have a mean (which is often called the "expected" statistic) and a standard deviation, which for sampling distributions we call the "standard error." In the next chapter, we explore how we can use the normal distribution to make inferences about the larger population given a random sample, its statistic, and the standard error of that statistic.

The χ^2 Distribution

In many cases, a statistical analysis makes use of the χ^2 distribution. This distribution is based on the standard normal distribution. While it is not directly used in this volume, it is an important stepping point on the way to the t and F distributions. This distribution is often useful for testing variances and sums of squares, as they are χ^2 distributed.[1]

---

[1] For example, the test of independence of two categorical variables is a χ^2 test. When looking at this test taught in most introductory statistics classes,

$$\sum_i \frac{(O_i - E_i)^2}{E_i},$$

where O_i is the observed cell count and E_i is the expected cell count. It is clear that we are summing squared deviations from expected values.

Figure 2.2 Areas of the normal distribution shaded with cumulative normal distribution function (Φ).

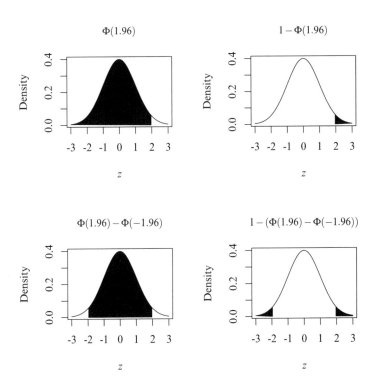

Essentially, the χ^2 distribution is the sum of independent squared Z distributions

$$\sum_{i=1}^{v} Z_i^2 = \chi_v^2. \qquad \text{(Equation 2.5)}$$

In this expression, v is the number of "degrees of freedom" of the variable.[2] In this case, it is the number of random Z variables we are adding

---

[2] Degrees of freedom is sometimes a difficult concept. Essentially, it is the number of values that can vary once we know a statistic about the value. For example, given the values 5 and 7, if we know the mean is 6, then we really only have 1 piece of information in the set {5,7}. This is because if I tell you that we have 2 numbers, 7 and an unknown number a, but that

14

Figure 2.3 Different χ^2 distributions with different degrees of freedom (v).

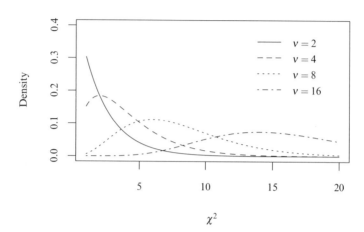

together. For example, we often use 1.96 as a critical value for two-tailed z tests when we set $\alpha = 0.05$. The square of one Z variable is a χ^2 with one degree of freedom. If you look in the back of any introductory statistics book and check the critical value of χ^2 with one degree of freedom, it is 3.84, which is the square of 1.96.

Since it is formed with squared Z variables, the value of χ^2 is always positive. Thus, it ranges from 0 to any positive number. As the degrees of freedom (v) increase, the distribution looks more normal.

Figure 2.3 presents various densities of χ^2 distributions. As with the normal distribution, we can use CDFs to find values at which we accumulate, for example, 95% of the distribution. This allows us to test whether random variables that are χ^2 distributed are highly unlikely if we expect the null hypothesis.

---

the mean is 6, we can figure out the unknown number with

$$6 = \frac{7+a}{2}$$

$$a = 6 \times 2 - 7 = 5.$$

This is why when working with contingency tables, the degrees of freedom for the independence test are the rows minus 1 times the columns minus 1.

Figure 2.4 Comparison of the normal and *t* distributions. *df*, degrees of freedom.

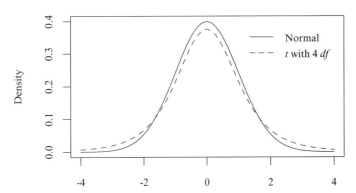

The *t* Distribution

The *t* distribution is the primary focus of this text. As we will see, it operates much like the standard normal distribution, except that it depends on the degrees of freedom (v) of the test. The formula and derivations for the *t* distribution are beyond the scope of this volume. The relationship between the *t* distribution with v degrees of freedom and the Z distribution is as follows (Rice, 2006):

$$t = \frac{Z}{\sqrt{\chi_v^2/v}} = \frac{Z}{\sqrt{\frac{1}{v}\sum_{i=1}^{v} Z_i^2}}.$$ (Equation 2.6)

The *t* distributions are "flatter" than the standard normal distribution (see Figure 2.4). As the degrees of freedom (v) increase, the *t* distributions approach the same shape as the normal distribution. As a result, the value for a cumulative percentage is different on the *t* distribution than it is on the standard normal distribution.

The *F* Distribution

The *F* distribution is often used for ratios of variances.[3] As such, it is based on χ^2s. A random *F* variable is the ratio of two χ^2s divided by

---

[3] It can also be used for correlations and multiple correlations as we will see in the last chapter. The reason for this is that correlations are standardized covariances.

their degrees of freedom (v):

$$F = \frac{\chi^2_{v_1}/v_1}{\chi^2_{v_2}/v_2}.$$
(Equation 2.7)

Thus, like the t distribution, the F distribution is based on the degrees of freedom. In the case of the F distribution, it is based on two degrees of freedom, one in the numerator and one in the denominator (v_1 and v_2, respectively). Figure 2.5 provides some example density curves.

F to t

In Chapters 6 and 7 we will make heavy use of the fact that when there is one degree of freedom in the numerator, the F distribution is the square of the t distribution (where $v = v_2$). Showing this fact brings together all the relationships we have reviewed thus far. First,

$$t = \frac{Z}{\sqrt{\chi^2_v/v}}$$

and

$$Z = \chi^2_1.$$

So, when $v_1 = 1$,

$$F = \frac{\chi^2_1/1}{\chi^2_{v_2}/v_2} = \frac{Z^2}{\chi^2_{v_2}/v_2},$$

Figure 2.5   Different F distributions with different degrees of freedom (v_1, v_2).

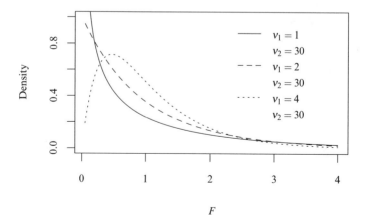

we can convert F into t by taking the square root,

$$\sqrt{F} = \frac{Z}{\sqrt{\chi_v^2/v}} = t.$$

Summary

In this chapter, the relationships between the standard normal, χ^2, t, and F distributions were explored. As test statistics are randomly distributed variables, having an understanding of their sampling distributions is important for understanding how to work with power analysis. The next chapter begins working with the standard normal distribution to lay the groundwork for working with other distributions for power analysis.

CHAPTER 3. GENERAL TOPICS IN HYPOTHESIS TESTING AND POWER ANALYSIS WHEN THE POPULATION STANDARD DEVIATION IS KNOWN: THE CASE OF TWO GROUP MEANS

This chapter is a review of population-based hypothesis testing (the z test) for the difference in means between two groups. We assume we know the population variance (σ^2) to keep things simple. In the subsequent chapters, we consider using samples (for which we have to estimate the population variance) to test differences between two groups, which affects the shape of the distributions that influence power.

The Difference in Means as a Normally Distributed Random Variable When the Population Standard Deviation Is Known

The beauty of inferential statistics is that the difference between two group means (as estimated from random samples) is also normally distributed when the population standard deviation is known. Consider two groups, group 0 (e.g., a control group) and group 1 (e.g., a treatment group). The population difference between two groups is defined as $\Delta = \mu_1 - \mu_0$. Its standard error (SE_Δ), which is the standard deviation of the distribution of differences between groups from many samples, is a function of the shared population standard deviation (σ) from which the groups are drawn, the proportion of observations in each group (P_0 and P_1, $P_0 + P_1 = 1, 1 - P_0 = P_1$), and the total number of observations (N):

$$SE_\Delta = \frac{\sigma}{\sqrt{NP_1(1 - P_1)}}. \qquad \text{(Equation 3.1)}$$

This expression, conceptually, means that the standard deviation of the sampling distribution is based on the standard deviation of the data divided by the square root of a function involving the sample size. The difference in group means for some variable y from a random sample is distributed normally such that

$$\bar{y}_1 - \bar{y}_0 \sim N(\Delta, SE_\Delta). \qquad \text{(Equation 3.2)}$$

Hypothesis Testing With the Difference Between Two Group Means When the Population Standard Deviation Is Known

The task with hypothesis testing is to determine the likelihood of observing the data given assumptions about the population. Qualitatively, this is a process in which we center the shape of the sampling distribution of the statistic on an assumed *null* hypothesis, and then find the probability that our data (or more extreme) came from such a sampling distribution. If the probability of observing the data on hand (or more extreme) is low, then we usually reject the null hypothesis as unlikely. The general logic is that either our data are wrong or the null hypothesis is wrong, and we use the process of quantifying the likelihood of our data (assuming the null hypothesis is true) to make our decision.

Hypothesis Testing Using Neyman and Pearson's Method

Hypothesis testing, as presented by the statisticians Neyman and Pearson (1933), is about making a decision between two hypotheses. The null hypothesis is generally noted as H_0, and the alternative hypothesis is noted as H_a. For example, we may state that the null hypothesis is that the means of group 0 and group 1 are the same,

$$H_0: \mu_1 = \mu_0,$$

which implies that the difference between the two group means, or Δ, is 0,

$$H_0: \mu_1 - \mu_0 = \Delta = 0.$$

When we are testing the difference in group means, this is typically the null hypothesis.

The alternative hypothesis (H_a) takes one of three forms when looking at the means of two groups. First, we can assert an alternative hypothesis that states that the mean of group 1 is greater than the mean of group 0,

$$H_a: \mu_1 > \mu_0,$$

which implies that $\Delta = \mu_1 - \mu_0$ is a positive value, or

$$H_a: \mu_1 - \mu_0 > 0,$$

or

$$H_a: \Delta > 0.$$

Second, we can assert an alternative hypothesis that states that the mean of group 1 is less than the mean of group 0,

$$H_a: \mu_1 < \mu_0,$$

which implies that $\Delta = \mu_1 - \mu_0$ is a negative value, or

$$H_a: \mu_1 - \mu_0 < 0,$$

or

$$H_a: \Delta < 0.$$

These first two alternative hypotheses are so-called one-tail hypotheses, the reason for which will be clear below. A final alternative hypothesis common to the examination of two group means is that the group means are just different, without a specification of direction,

$$H_a: \mu_1 \neq \mu_0,$$

which implies that the difference $\Delta = \mu_1 - \mu_0$ is nonzero,

$$H_a: \mu_1 - \mu_0 \neq 0,$$

or

$$H_a: \Delta \neq 0.$$

This is a so-called two-tailed test, and why it is called a two-tailed test will be explained below.

The decision with hypothesis testing is this: Should the null hypothesis (H_0) be rejected in favor of the alternative hypothesis (H_a)? A general way to think about rejecting the null hypothesis is to state that the observed data are unlikely to have occurred if the null were true, since the observed Δ is so different from the value stated in the null hypothesis relative to the sampling distribution.

One way to do this is to make the assumption that Δ is the value stated in the null hypothesis (typically, the null hypothesis states that $\Delta = 0$), and then draw a sampling distribution that is normally distributed with a mean of $\Delta = 0$ and a standard deviation of SE_Δ. Remember, SE_Δ is based on the known population standard deviation, the sample size, and the proportion of observations in each group (Equation 3.1). To put it another way, our sample size and known standard deviation allow us to draw a normal curve centered on the value of Δ assumed by the null hypothesis.

We then consider how likely it is to have observed our Δ compared with the null hypothesis. This approach, however, is not standardized and depends the ability to plot sampling distributions in the unit of the outcome being considered. While this is possible, it is difficult to do and does not lend itself well to general rules of thumb.

Another possibility is to perform the same procedure, but instead of using Δ and its standard error to draw a sampling distribution, we use Δ and its standard error to calculate a standard test statistic and then use the standard normal distribution to consider the likelihood of the data in light of the null hypothesis. The test statistic is implied in the z-scores used in Equations 2.2, 2.3, and 2.4, and is

$$z = \frac{\Delta - \Delta_0}{SE_\Delta}, \qquad \text{(Equation 3.3)}$$

where Δ_0 is the difference assumed by the null hypothesis (which is typically 0). With this test in hand, we can return to the standard normal distribution in Figure 2.1 to set up our areas of rejection that relate to Type I error.

We now return to the types of errors that were introduced in Chapter 1. The first type of error is Type I, which is the chance of rejecting the null hypothesis when in fact it is true. We note this error as α, and typically consider $\alpha = 0.05$ as an acceptable chance of Type I error.

We use the Type I error (α) rate with the sampling distribution under the null hypothesis to create regions of rejection based on the specification of the alternative hypothesis. These rejection regions are simply areas of the sampling distribution of the null hypothesis that encompass the proportion of area under the distribution curve that represents an acceptable risk of error. The value of z where the rejection region begins is the critical value, and if the test statistic (Equation 3.3) exceeds this value, then the null hypothesis is rejected.

Figure 3.1 presents these critical regions for each of the three types of alternative hypotheses that set $\alpha = 0.05$. As you can see, the first two plots only shade one of the tails with 5% ($\alpha = 0.05$) of the area (which is why they are called "one-tailed" tests). The shading for the first plot, the "greater-than" alternative hypothesis that $\mu_1 > \mu_0$, starts at $z = 1.64$. This means that if Equation 3.3 is greater than 1.64, we reject the null hypothesis in favor of the alternative hypothesis. A similar logic is used for the alternative hypothesis used in the second plot of Figure 3.1, that $\mu_1 < \mu_0$, except here the test statistic must be less than -1.64 to reject the null hypothesis.

Finally, if the alternative hypothesis is that the means are simply different, that $\mu_1 \neq \mu_0$, we split the 5% of the shaded area into both tails, which is why they are "two-tailed" tests. Since we now shade the

22

Figure 3.1 Null z distribution with one- and two-tailed critical regions shaded ($\alpha = 0.05$).

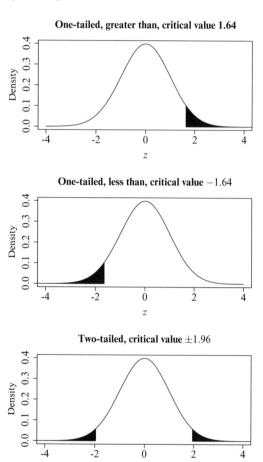

tails with only 2.5%, the critical values are more extreme, $z = -1.96$ and $z = 1.96$. If the test statistic (Equation 3.3) exceeds 1.96 in either direction, we reject the null hypothesis. Of course, as we change α to another value, for example, $\alpha = 0.01$, the percentage of the sampling distribution that becomes the region of rejection changes, and thus the critical values change. Critical values for various levels of α and alternative hypotheses are found in almost every introduction to statistics textbook.

What Is a p Value?

Most statistical software programs do not use the Neyman and Pearson method of hypothesis testing directly. Instead, the software finds the two-tailed area of the sampling distribution that exceeds the magnitude of a test statistic, assuming that the null hypothesis is $\Delta_0 = 0$. For example, if the test statistic (Equation 3.3) were computed to be $z = 3$, the software uses the equivalent of one minus Equation 2.4 to compute a probability that the data's test result (or a result that is more extreme) would occur assuming a null hypothesis of $\Delta = 0$:

$$p(z) = 1 - (\Phi(z) - \Phi(-z))$$
$$p(z) = 1 - (\Phi(3) - \Phi(-3))$$
$$p(z) = 0.003$$

Another way to express the two-tailed p value is

$$p(z) = 2 \times \left(1 - \Phi\left(\frac{|\Delta|}{SE_\Delta}\right)\right), \qquad \text{(Equation 3.4)}$$

which for the one-tailed test would be,

$$p(z) = 1 - \Phi\left(\frac{|\Delta|}{SE_\Delta}\right). \qquad \text{(Equation 3.5)}$$

The idea behind the "test" that uses the p value is if $p(z)$ is less than α, then we consider the difference to be "significant."

It is important to realize that p values are not the probability that the data per se occur if the null hypothesis were true. The data are a single point along a continuous distribution. Probabilities from a continuous distribution require a *range*, and the probably of a single point is 0. That is why we talk about the p value as the chance that "the data's test result or a result that is more extreme" would occur if the null hypothesis were in fact true.

What Is Useful for Power Analysis?

As we will see below, power analysis is essentially the manipulation of areas under curves given various ranges of values. Thus, p values are less useful for power analyses since we never know the exact p value a future dataset will generate. Instead, we can prespecify a level of α, which generates critical values, which allow us to find the probabilities of ranges along normal distributions (and other distributions when we have to estimate the population standard deviation, σ).

Power Analysis for Testing the Difference Between Two Group Means When the Population Standard Deviation Is Known

Now that we have established how to think about hypothesis testing using the normal distribution, we are ready to think about power analysis. Power analysis is tied to Type II error (β), or the chance that we accept the null hypothesis when in fact the alternative is true. One way to think about Type II error is to consider the chance that our test statistic falls before the critical value specified by α, analogous to the second simulation in Chapter 1 (see Figure 1.2). For example, if our alternative hypothesis states that $\Delta > 0$, and our one-sided test sets $\alpha = 0.05$, our critical value of z is 1.64. Before collecting the data, Type II error asks what is the chance that our test statistic (Equation 3.3) winds up being less than 1.64, even though the alternative hypothesis is true?

To answer this question, we suppose some expected test result and then draw a sampling distribution around that expected test result. Noting where the critical value falls, Type II error (β) for a one-tailed test is simply the area before the right critical value (if the alternative is positive) or after the left critical value (if the alternative is negative). Type II error (β) for a two-tailed test is the area between the critical values. The power of the test is then the area outside this region, or $1 - \beta$. Thus, power analysis is about imposing a sampling distribution that assumes the alternative hypothesis is true (the *alternative* or noncentral distribution) onto the distribution that assumes the null hypothesis is true (the *null* or central distribution).

In order to find areas under the curve between points of the the alternative distribution, we need to know the mean and standard deviation of the alternative distribution. Using the standard normal distribution, we can use the same standard deviation as the null curve for the alternative curve (which is 1). However, while the mean of the null distribution is 0 (because the null assumes that $\Delta = 0$), we need to find an appropriate average for the alternative distribution. The mean for the alternative distribution is the expected test statistic, or what we expect Equation 3.3 to be if the alternative were true. We generally note this parameter to be λ and call it the noncentrality parameter. Rearranging Equation 3.3 slightly, using Δ and Equation 3.1, gives a general expression for the noncentrality parameter for a standard normally distributed test statistic:

$$\lambda = \frac{|\Delta|}{\sigma} \sqrt{NP_1(1 - P_1)}. \qquad \text{(Equation 3.6)}$$

Note that in this volume we will consider the difference between means in absolute values. This allows us to organize our analyses around only positive critical values.

Once we have a value for λ, we can take advantage of the fact that a variable Z, which is distributed standard normal, minus a noncentrality value λ is normally distributed (i.e., $Z - \lambda \sim N(0, 1)$). The way we use this fact is to use the CDF of the standard normal distribution (Φ) to find the area before a critical value of $z_{1-\alpha}$ given a level of α to estimate the Type II error of a one-tailed test,

$$\beta = \Phi(z_{1-\alpha} - \lambda), \qquad \text{(Equation 3.7)}$$

or the area between critical values to estimate the Type II error for two-tailed tests,

$$\beta = \Phi(z_{1-\alpha/2} - \lambda) - \Phi(z_{\alpha/2} - \lambda). \qquad \text{(Equation 3.8)}$$

With Type II error (β) in hand, power is then simply $1 - \beta$.

Example

Suppose that a difference in means between two groups for some standardized test is expected to be $\Delta = 25$ from a sample of 100 observations where the population standard deviation is $\sigma = 75$. Furthermore, suppose a balanced design where 50 observations are in each group, so $P_1 = \frac{50}{100} = 0.5$. Using Equation 3.6, we can calculate λ as

$$\lambda = \frac{25}{75} \sqrt{100 \times 0.5 \times (1 - 0.5)} = 1.67.$$

If we were to conduct a one-tailed test with $\alpha = 0.01$, then the critical value of the standard normal distribution is 2.326, or the $1 - \alpha = 1 - 0.01 = 0.99$ quantile value of the z distribution is $z_{1-\alpha} = z_{0.99} = 2.326$. We then find the Type II error by finding the area of the standard normal distribution before $z_{1-\alpha} - \lambda = 2.326 - 1.667 = 0.659$, or $\beta = \Phi(0.659) = 0.745$. This means that power is about $1 - \beta = 1 - 0.745 = 0.255$. The one-tailed example is featured in Figure 3.2. As you can see, the gray shaded area is $\beta \times 100 = 75\%$ of the alternative distribution, leaving the area beyond the right critical value (where the black shading starts) as the power of the test.

If we were to use the same value of λ and conduct a two-tailed test with $\alpha = 0.01$, then the critical value of z is 2.576, or the $1 - \alpha/2 = 1 - 0.01/2 = 0.995$ quantile value of the z distribution is

Figure 3.2 The standard normally distributed null distribution (solid line) with one tail shaded corresponding to $\alpha = 0.01$ where the critical value $z_{\alpha/2} = 2.326$, and the alternative distribution (dashed line) with a noncentrality parameter $\lambda = 1.667$ and $\beta = 0.745$.

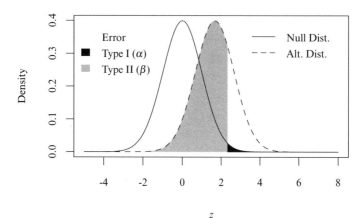

$z_{1-\alpha/2} = z_{0.995} = 2.576$. We find the Type II error by finding the area of the normal distribution between $z_{1-\alpha/2} - \lambda$ and $z_{\alpha/2} - \lambda$, which is $\beta = \Phi(2.576 - 1.667) - \Phi(-2.576 - 1.667) = 0.818$. This means that power is about $1 - \beta = 1 - 0.818 = 0.182$. The two-tailed example is featured in Figure 3.3. As you can see, the gray shaded area is $\beta \times 100 = 82\%$ of the alternative distribution, leaving the area beyond the left and right critical values (where the black shading starts) as the power of the test.

Useful Relationships

Given a standard normal distribution, we can use Equations 3.7 and 3.8 to calculate useful quantities. For example, consider the Type II error for a one-tailed test:

$$\beta = \Phi(z_{1-\alpha} - \lambda).$$

Here, β is an area of the normal curve. Thus, we can cancel the Φ on both sides of the equation to reveal that

$$z_\beta = z_{1-\alpha} - \lambda,$$

which is reorganized as

$$\lambda = z_{1-\alpha} - z_\beta.$$

Figure 3.3 The standard normally distributed null distribution (solid line) with two tails shaded corresponding to $\alpha = 0.01$ where the critical value $z_{\alpha/2} = \pm 2.576$, and the alternative distribution (dashed line) with a noncentrality parameter $\lambda = 1.667$ and $\beta = 0.818$.

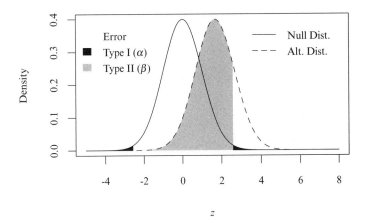

This means that in the context of the standard normal test, the expected test statistic is a function of two quantiles (at points $1 - \alpha$ and β) of the standard normal distribution. The two-tailed test is a little more difficult, because we are adding two CDFs to calculate β. However, the second term, $\Phi\left(z_{\alpha/2} - \lambda\right)$, is usually very small (it is the area of the test distribution before the left critical value). In the example, the value was 0.00001. If we ignore that portion, we can use the same method for the one-tailed test to find

$$\lambda \approx z_{1-\alpha/2} - z_\beta$$

for the two-tailed test. Thus, we can generalize this relationship with the notation of the positive value of $z_{critical}$ based on whether the test is one- or two-tailed:

$$\lambda \approx z_{critical} - z_\beta. \qquad \text{(Equation 3.9)}$$

If $\alpha = 0.05$, the value of $z_{critical} = 1.96$ for two-tailed tests and $z_{critical} = 1.68$ for one-tailed tests. For power of 0.8, $z_\beta = z_{0.2} = -0.84$, for power of 0.9, $z_\beta = z_{0.1} = -1.28$. For example, the expected test statistic (λ) for an α two-tailed test with power of 0.8 is approximately $1.96 - (-0.84) = 2.8$, and that number is often used in different types of power analyses as shown below.

Distribution Quantiles

The values of $z_{critical}$ and z_β are quantiles of the standard normal distribution. A quantile is the inverse of a CDF. It is the value of z given a cumulative probability. Finding the quantiles of the normal distribution is possible with a computer program. For ease of using this text, I present important quantiles of the standard normal distribution in the last row of Table 4 in the Appendix.

The first set of columns in Appendix Table 4 are useful in finding quantiles of the normal distribution associated with values of β. For example, for a Type II error of 0.1, use $z_\beta = z_{0.1} = -1.282$. The columns to the right are useful for finding the positive critical values associated with the type of test. For example, if $\alpha = 0.05$ for a one-tailed test, use $z_{1-\alpha} = z_{0.95} = 1.645$. For a two-tailed test, use $z_{1-\alpha/2} = z_{0.975} = 1.96$.

Scale-Free Parameters

Up until this point, almost all the components of our power analysis have been in the units of the outcome. The values of μ_1, μ_0, Δ, and σ have all been in the units of the variable in question, whether it be a standardized test, body mass index, or Galaga scores. The actual difference in values in the means between groups is itself difficult to anticipate, and in samples the variance is also difficult to anticipate. Power analysis becomes much easier if we deal with so-called "scale-free" parameters. Our first scale-free parameter to be introduced in this volume is the effect size. Cohen's d is an effect size that standardizes the difference between group means, Δ, by dividing it by the population standard deviation, σ (Cohen, 1988). This metric can be expressed with the lowercase Greek letter δ, and is noted as

$$\delta = \frac{\mu_1 - \mu_0}{\sigma} = \frac{\Delta}{\sigma}, \qquad \text{(Equation 3.10)}$$

which is the first term in Equation 3.6.

The meaning of δ is relatively clear: It is the difference in means in standard deviation units. Thus, $\delta = 0.5$ indicates that one group is half a standard deviation greater, on average, than another group. The purpose of using δ instead of $\frac{\Delta}{\sigma}$ in planning studies is twofold. First, it allows researchers to think about the results of their future study in more general terms; what is the general difference between groups in standard units?

Second, and probably more important, is that effect sizes allow researchers to use prior research to manage expectations (see Chapter 8), even though they use different metrics. For example, suppose researchers are planning a study on how child savings accounts for college influence the financial literacy of parents. A previous study may use scale A, whereas the current researchers may plan to use scale B. If scale A ranges from 1 to 100, it is hardly possible to translate that difference into scale B, which ranges from 0 to 7. However, if the previous research published the means and standard deviations for treatment and control groups, it would be possible to determine an effect size (more on this later). With an effect size in hand, researchers using scale B will have some insight about the expected effect in their study.

Thus, we can express the noncentrality parameter using the effect size as follows:

$$\lambda = \underbrace{\delta}_{\text{Effect size}} \underbrace{\sqrt{NP_1(1-P_1)}}_{\text{Sample size}}, \qquad \text{(Equation 3.11)}$$

which allows for a power analysis that has no information about the scaled parameters in the future study.

Balanced or Unbalanced?

Another issue often encountered in planning studies is *balance*. A balanced study is one in which each group has the same number of cases, whereas unbalanced studies have differing numbers of cases in each group. Balance was once a crucial consideration in studies, especially before modern computing, because it made the calculations easier. Today, it is rarely the case that the final data researchers work with will actually be balanced; but it may be close.

However, in planning studies, balance is an important consideration for several reasons. First, holding total sample size constant, a balanced sample has more power than an unbalanced sample. Second, as you will see, the noncentrality parameter and the calculations that employ it are easier to handle with a balanced sample. Third, sometimes balance may not be palatable: We may want to have a smaller control group to maximize the number of treatments applied to the study population.

Noncentrality Parameter Without Balance

We have already seen the noncentrality parameter for the unbalanced case (Equation 3.11). In this case, group 1 has a sample size of n_1 and group 0 has a sample size of n_0, the total sample size is $n_1 + n_0 = N$, and the proportion of observations in group 1 is $\frac{n_1}{N} = P_1$. Thus, the calculations in the unbalanced case need information about the effect size, δ, the *total sample*, N, and the proportion of observations in one of the groups (I prefer group 1 P_1, but it makes no difference).

Noncentrality Parameter With Balance

The noncentrality parameter simplifies somewhat when $n_1 = n_0 = n$. In this case, $P_1 = (1 - P_1) = 0.5$ and the second term in the noncentrality parameter (Equation 3.11) is

$$\sqrt{N \times 0.5^2} = \sqrt{N \times 0.25} = \sqrt{\frac{N}{4}} = \sqrt{\frac{n}{2}}.$$

This makes the noncentrality parameter in the balanced case

$$\lambda = \underbrace{\delta}_{\text{Effect size}} \underbrace{\sqrt{\frac{n}{2}}}_{\text{Sample size}}. \qquad \text{(Equation 3.12)}$$

Thus, the calculations in the balanced case need only information about the effect size, δ, and the *sample size in each group, n*.

Types of Power Analyses

Three types of questions are usually posed when planning the sample design of a study:

- **Finding power a priori:** What will be the power for a given α, effect size, and sample size?

- **Finding the necessary sample size:** What will the sample size need to be for a given level of power $(1 - \beta)$, α, and effect size?

- **Finding the minimum detectable effect size:** What is the minimum detectable effect size for a given level of power $(1 - \beta)$, α, and sample size?

Finding Power A Priori

Much of this volume so far as been about finding power given an effect size, sample size, and level of α. To briefly recap, once we have an expected effect size δ as defined in Equation 3.10, the total sample size N, proportion of cases in one group P, and α, we can find the Type II error (β) for a one-tailed test using the CDF of the standard normal distribution,

$$\beta = \Phi(z_{1-\alpha} - \lambda) = \Phi\left(z_{1-\alpha} - \delta\sqrt{NP_1(1-P_1)}\right),$$

and a similar process for a two-tailed test,

$$\beta = \Phi(z_{1-\alpha/2} - \lambda) - \Phi(z_{\alpha/2} - \lambda),$$

$$\beta = \Phi\left(z_{1-\alpha/2} - \delta\sqrt{NP_1(1-P_1)}\right) - \Phi\left(z_{\alpha/2} - \delta\sqrt{NP_1(1-P_1)}\right),$$

where z_a is the quantile of the standard normal distribution at point a.

When the data are expected to be balanced, these formulas simplify to the following for a one-tailed test,

$$\beta = \Phi(z_{1-\alpha} - \lambda) = \Phi\left(z_{1-\alpha} - \delta\sqrt{\frac{n}{2}}\right),$$

and to the following for a two-tailed test,

$$\beta = \Phi(z_{1-\alpha/2} - \lambda) - \Phi(z_{\alpha/2} - \lambda),$$

$$\beta = \Phi\left(z_{1-\alpha/2} - \delta\sqrt{\frac{n}{2}}\right) - \Phi\left(z_{\alpha/2} - \delta\sqrt{\frac{n}{2}}\right).$$

These calculations require either a computer with the standard normal CDF function or a table of values. In any case, once the value of β is determined, the power of the test is $1 - \beta$.

Finding the Necessary Sample Size

To find the necessary sample size, we return to the useful relationship presented above, namely that $\lambda \approx z_{critical} - z_\beta$. With this relationship, and unpacking the noncentrality parameter λ, we can find the sample size necessary to detect a given effect size for a given level of α, which drives $z_{critical}$, and power, which drives z_β.

Unbalanced Case

In the unbalanced case, the noncentrality parameter (Equation 3.11) for testing the difference between two groups with a known standard deviation is

$$\lambda = \delta\sqrt{NP_1(1-P_1)},$$

and so we can express all the ingredients for this calculation as

$$\delta \sqrt{NP_1 (1 - P_1)} \approx z_{critical} - z_\beta.$$

We can rearrange this to isolate N, which leaves us with the formula

$$N \approx \frac{\left(z_{critical} - z_\beta\right)^2}{(P_1 (1 - P_1)) \delta^2}. \qquad \text{(Equation 3.13)}$$

Looking at this expression, we can see that holding constant the power and significance level (the numerator), the sample size will increase as P_1 deviates from 50%, and will decrease as the effect size (δ) increases.

Example

For example, suppose we were going to perform a two-tailed test with $\alpha = 0.05$ and thus $z_{critical} = z_{1-\alpha/2} = z_{0.975} = 1.96$, and wished to find the sample size for power of 0.8, so $\beta = 1 - 0.8 = 0.2$ and thus $z_\beta = z_{0.2} = -0.842$, with an effect size of $\delta = 0.3$, where a quarter of the sample was in group 1 and thus $P_1 = 0.25$. Putting these values into Equation 3.13, we arrive at

$$N \approx \frac{\left(z_{critical} - z_\beta\right)^2}{(P_1 (1 - P_1)) \delta^2} \approx \frac{(1.96 - (-0.842))^2}{(0.25 (1 - .025)) 0.3^2} \approx 465.26,$$

or about 466 observations (we always round up).

Balanced Case

If the observations are balanced, that is, $n_1 = n_0 = n$ and $2 \times n = N$, the noncentrality parameter reduces to Equation 3.12,

$$\lambda = \delta \sqrt{\frac{n}{2}} \approx \left(z_{critical} - z_\beta\right),$$

which can be arranged to find

$$n \approx \frac{2 \left(z_{critical} - z_\beta\right)^2}{\delta^2}. \qquad \text{(Equation 3.14)}$$

As in the unbalanced case, we find again this expression indicating that the sample size will decrease as the effect size (δ) in the denominator increases.

Example

For example, suppose we were going to perform a two-tailed test with $\alpha = 0.05$ and thus $z_{critical} = z_{1-\alpha/2} = z_{0.975} = 1.96$, and wished to find the sample size for power of 0.8, so $\beta = 1 - 0.8 = 0.2$ and thus $z_\beta = z_{0.2} = -0.842$, with an effect size of $\delta = 0.3$. Putting these values into Equation 3.14, we arrive at

$$n \approx \frac{2\left(z_{critical} - z_\beta\right)^2}{\delta^2} \approx \frac{2\left(1.96 - (-0.842)\right)^2}{0.3^2} \approx 174.47$$

or about 175 cases per group for a total of $2n = N = 350$. Notice that all the parameters in this example were the same as in the unbalanced case, and we arrived at a smaller value of N.

Finding the Minimum Detectable Effect Size

The minimum detectable effect size (Bloom, 1995) is a metric designed to summarize the sensitivity of a given sample. It is the smallest effect size (δ) that can be detected at a given level of power, assuming a sample size and level of α. We again return to the useful relationship presented above, namely that $\lambda \approx z_{critical} - z_\beta$. With this relationship, and unpacking the noncentrality parameter λ, we can find the effect size necessary to result in a significant test for a given sample size, with a given level of α, which drives $z_{critical}$, and power, which drives z_β.

Unbalanced Case

Rearranging the noncentrality parameter (Equation 3.11) and $\left(z_{critical} - z_\beta\right)$, we can isolate the minimum detectable effect size

$$\delta_m \approx \frac{z_{critical} - z_\beta}{\sqrt{NP_1(1 - P_1)}}. \qquad \text{(Equation 3.15)}$$

This expression indicates that the detectable effect will decrease as the sample size (N) increases, but will increase as the sample becomes more unbalanced (i.e., P_1 moves away from 0.5).

Example

For example, suppose we were going to perform a two-tailed test with $\alpha = 0.05$ and thus $z_{critical} = z_{1-\alpha/2} = z_{0.975} = 1.96$, and we wished to find the sample size for power of 0.8, so $\beta = 1 - 0.8 = 0.2$ and thus $z_\beta = z_{0.2} = -0.842$, with a sample size of $N = 500$, where a quarter of the sample was in group 1 and thus $P_1 = 0.25$. Putting these values into Equation 3.15, we arrive at

$$\delta_m \approx \frac{z_{critical} - z_\beta}{\sqrt{NP_1(1 - P_1)}} \approx \frac{1.96 - (-0.842)}{\sqrt{500 \times 0.25(1 - 0.25)}} \approx 0.289.$$

This effect size is a little smaller than the previous example of 0.3 because we increased the sample from 466 to 500.

Balanced Case

If the observations are balanced, i.e., $n_1 = n_0 = n$ and $2 \times n = N$, the noncentrality parameter can be rearranged to find

$$\delta_m \approx (z_{critical} - z_\beta)\sqrt{\frac{2}{n}}. \qquad \text{(Equation 3.16)}$$

This expression indicates that the minimum detectable effect will decrease as the sample size increases.

Example

For example, suppose we were going to perform a two-tailed test with $\alpha = 0.05$ and thus $z_{critical} = 1.96$, and wished to find the sample size for power of 0.8, so $\beta = 1 - 0.8 = 0.2$ thus $z_\beta = 0.842$, with a sample size of $n = 175$ per group. Putting these values into Equation 3.16, we arrive at

$$\delta_m \approx (z_{critical} - z_\beta)\sqrt{\frac{2}{n}} \approx (1.96 - (-0.842))\sqrt{\frac{2}{175}} \approx 0.300$$

or an effect size of 0.3, which is associated with 175 cases per group as above.

Power Tables

The focus of this volume is on calculating results using formulas, which in turn will inform the use of software. Before software was widely available for power analysis, many researchers relied on books of tables such as Cohen's seminal volume (1988). These tables allowed the researcher to perform all three types of power analyses (with some effort). For example, examine Table 3.1, which is a reproduction of a similar table in Cohen's book (1988).

Power tables are typically organized around the three key pieces of information: sample size, effect size, and power. Separate tables are produced around the other assumptions, such as the number of tails and Type I error (α). With tables such as Table 3.1, users can move across rows that identify sample sizes, and columns, which identify effect sizes, to find the power in a given cell. For example, looking at Table 3.1, we can see that for a sample size of 58 units in each group (last row) and an effect size of 0.4 (second to last column), the power of that study for a two-tailed test with $\alpha = 0.05$ is 0.57 (power tables typically do not print the decimal point).

Table 3.1 Portion of Power Table 2.3.5 for Two-Tailed Tests With $\alpha = 0.05$ From Cohen (1988, p. 37)

n	δ				
	0.1	0.2	0.3	0.4	0.5
50	8	17	32	51	70
52	8	17	33	52	71
54	8	18	34	54	73
56	8	18	35	55	75
58	8	19	36	57	76

Note. Small differences in power values are due to the approximation used in Cohen (1988) compared with the software used by the present author.

Summary

In this chapter, we focused on how power analysis works from the point of view of understanding the sampling distributions under two scenarios: one in which the null hypothesis the true (the central distribution) and the other in which the alternative is true (the noncentral or alternative distribution). If the population standard deviation is known, we can use the standard normal distribution and its quantiles to perform a power analysis. I also introduced the idea of the effect size and the need for scale-free parameters. In the next chapter, we move to the more common scenarios in which the population standard deviation is not known and must be estimated. This means that we must use the t distribution (Student, 1908), which depends on degrees of freedom, which in turn depends on the sample size. Since sample size is such an important aspect of power, we will find that this complicates matters.

CHAPTER 4. THE DIFFERENCE BETWEEN TWO GROUPS IN SIMPLE RANDOM SAMPLES WHERE THE POPULATION STANDARD DEVIATION MUST BE ESTIMATED

In this chapter, we revisit the test about differences between group means, but now we acknowledge that the population standard deviation is unknown and must be estimated. This changes the distribution that we must use for the test and to determine power. We now must use the t distribution instead of the standard normal distribution (Student, 1908). Since the t distribution depends on the sample size, power analysis is less straightforward. In this chapter, we focus on the least squares regression method for testing for group mean differences. Doing this lays the groundwork for adding covariates to the model and also for clustered designs.

As a working example of the balanced case, we will look at data from an experiment about breakfast practices and weight gain. This study, led by Dhurandhar and colleagues (2014), was titled "The Effectiveness of Breakfast Recommendations on Weight Loss" and was a randomized controlled trial to assess how different approaches to eating breakfast affect weight loss. This study randomly assigned treatment conditions to a study sample of obese,[1] but otherwise healthy, individuals. The treatment conditions included a group that was told to eat breakfast, a group that was told not to eat breakfast, and a control group that was only given nutritional guidance. The study found that while eating habits did vary between treatment groups, there was no statistical difference in the participants' body mass index (BMI). BMI is defined as weight in kilograms divided by the square of one's height in meters and is expressed in units of kg/m^2.

For the purposes of this volume, we compare only select cases who ate breakfast with those who did not eat breakfast. We use as our outcome the study participants' BMI measured at the end of the study period. It should also be noted that this study is registered at clinicaltrials.gov as NCT01781780, and the documentation includes a power analysis.

---

[1] As you will see in Table 4.1, the average BMI is about 33. The cutoffs for BMIs are 25 to 29.999, overweight; ≥ 30, obese.

Table 4.1   Summary Statistics of Body Mass Index Scores After Treatment

	N	Mean	Standard Deviation
No breakfast (control)	25.000	35.340	5.966
Breakfast (treatment)	25.000	31.459	5.476
Total sample	50.000	33.400	5.997

The important information from the subsample for our purposes (means, standard deviations, and sample sizes) is presented in Table 4.1. The raw data used for this analysis are located in the Appendix. Although only a tiny fraction of the actual data is used in this volume, the full public use data file is available at ICPSR[2] (study number 36174).

Data-Generating Process

We begin with what experimentalists call the experimental design model equation (Kirk, 1995). Other fields think of this as the "data-generating process." The model we assume is

$$y_{ij} = \mu + \tau_j + e_{ij},\qquad\text{(Equation 4.1)}$$

where y_{ij} is the ith observation in treatment group j of the outcome from a population with variance σ^2, μ is the grand mean of the outcome among all observations, τ_j is the treatment effect[3] for group j (with the constraint that $\sum_j \tau_j = 0$), and e_{ij} is the error associated with the ith observation in treatment group j. For each treatment group, there are $i = \{1,2,3,\ldots,n_j\}$ observations, and there can be $j = \{1,2,\ldots,p\}$ groups. In this text, we are concerned with the case of two groups, or $p = 2$. Also, this text will subscript the treatment group with $j = 1$ and the control group with $j = 0$ to be consistent with the coding of the treatment indicator. Thus, the sample size for group $j = 0$ is n_0, and the sample size for group $j = 1$ is n_1. The total sample size is then $N = n_0 + n_1$.

---

[2] Inter-university Consortium for Political and Social Research, www.icpsr.umich.edu.
[3] Kirk, and many others, use α_j to denote the treatment effect, but in the context of an introductory book on power, where the Type I error level of α plays such a large role, I find this confusing. I therefore use τ_j.

Testing the Difference Between Group Means With Samples

Sampling distributions of random variables, like the difference between group means, are formed with means and variances. In this section, we explore the difference in means and the sampling variance of the difference in means from a regression point of view. This is done for the *unbalanced* case, where each group has a different number of observations (i.e., $n_0 \neq n_1$), and the *balanced case*, where each group has the same number of observations (i.e., $n_0 = n_1 = n$).

The grand mean (μ) of the outcome is estimated from the sum of all observations from all groups divided by the total sample size

$$\bar{y} = \frac{\sum_j \sum_i y_{ij}}{N}. \qquad \text{(Equation 4.2)}$$

The estimation for the mean for group j, μ_j, is estimated with

$$\bar{y}_j = \frac{\sum_i^{n_j} y_{ij}}{n_j}. \qquad \text{(Equation 4.3)}$$

The treatment effect τ_j is equal to the difference between the group mean and the grand mean, that is, $\tau_j = \mu_j - \mu$. Since we assume that there are only p treatment options in the population, we will view the effects as fixed and not a random selection from a population of treatments. If we were dealing with a random selection of possible treatments, the analysis would be different. Finally, e_{ij} is the within-group residual error, which is the difference between the ith observation in group j and its group mean, $e_{ij} = y_{ij} - \mu_j$. Since the treatment effect is the only covariate, the variance of the residuals is the variance of the population without treatments introduced, so the variance of e_{ij} is σ^2.

As you will see below, similar to Chapter 3, the sampling variance of the difference in means is distributed with a variance of Equation 4.15 that depends on an estimate of the population variance σ^2, which is unknown. This quantity must be estimated from the data. One option could be using the variance as it is typically estimated for survey data:

$$\tilde{\sigma}^2 = \frac{\sum_j \sum_i (y_{ij} - \bar{y})^2}{N - 1}. \qquad \text{(Equation 4.4)}$$

This is not a good estimate of the population variance since a treatment was introduced that (hopefully) introduces variance in the data. A better estimate replaces the grand mean, $\bar{y}$, with the group mean, $\bar{y}_j$, and adjusts the degrees of freedom in the denominator:

$$\hat{\sigma}^2 = \frac{\sum_j \sum_i (y_{ij} - \bar{y}_j)^2}{N - p}. \qquad \text{(Equation 4.5)}$$

This produces what introductory texts call the pooled variance, which is typically defined as the weighted average of the variances of the outcome for each group. In the case of two groups, this is

$$\hat{\sigma}^2 = \frac{(n_1 - 1)\hat{\sigma}_1^2 + (n_0 - 1)\hat{\sigma}_0^2}{n_1 + n_0 - 2}, \qquad \text{(Equation 4.6)}$$

where the variance of the treatment group is defined as

$$\hat{\sigma}_1^2 = \frac{\sum_i (y_{i1} - \bar{y}_1)^2}{n_1 - 1}$$

and the variance of the control group is defined as

$$\hat{\sigma}_0^2 = \frac{\sum_i (y_{i0} - \bar{y}_0)^2}{n_0 - 1}.$$

The Regression Model

Simple bivariate regression employs the least squares criteria to fit the best summary line to describe the relationship between two variables. We can think of the model for a regression analysis as [4]

$$y_{ij} = \gamma_0 + \gamma_1 T_{ij} + e_{ij}, \qquad \text{(Equation 4.7)}$$

where y is the linear outcome of interest and T is the explanatory treatment variable.[5]

To estimate the regression slope γ_1, most introductory statistics texts provide the equivalent to the following formula:

$$\hat{\gamma}_1 = \frac{\sum_j \sum_i (y_{ij} - \bar{y})(T_{ij} - \bar{T})}{\sum_j \sum_i (T_{ij} - \bar{T})^2}, \qquad \text{(Equation 4.8)}$$

---

[4] To avoid confusion with the Type II error parameter β, I use γ for regression slopes.
[5] I am using T as the predictor, and not x, to avoid confusion with the covariate we will consider later.

where $\bar{T}$ is the mean of the treatment indicator. In the case of two groups, a common way to code the predictor is as a dichotomous or "dummy" variable:

$$T_{ij} = \begin{cases} 0 & \text{if Control,} \\ 1 & \text{if Treatment.} \end{cases} \qquad \text{(Equation 4.9)}$$

With this coding, the mean of T is simply

$$\bar{T} = \frac{\Sigma_j \Sigma_i T_{ij}}{N} = \frac{n_1}{N} = P_1. \qquad \text{(Equation 4.10)}$$

When T is a dichotomous indicator, we will see below that the slope $\hat{\gamma}_1$ is an estimate of the difference between the mean of cases where $T = 1$ and the mean of cases where $T = 0$, $\gamma_1 = \bar{y}_1 - \bar{y}_0$.

The intercept of a bivariate regression model is also straightforward to estimate,

$$\hat{\gamma}_0 = \bar{y} - \hat{\gamma}_1 \bar{T}, \qquad \text{(Equation 4.11)}$$

and we see below that this quantity is equal to the mean of group 0.

The estimated sampling variance of the slope, γ_1, in a bivariate regression is often presented as

$$\text{vâr}\{\hat{\gamma}_1\} = \frac{\hat{\sigma}_e^2}{\Sigma_j \Sigma_i (T_{ij} - \bar{T})^2} \qquad \text{(Equation 4.12)}$$

where the mean squared error (MSE or σ_e^2) of a bivariate regression is estimated as

$$\hat{\sigma}_e^2 = \frac{\Sigma_j \Sigma_i \hat{e}_{ij}^2}{N - 2}. \qquad \text{(Equation 4.13)}$$

In the regression framework, e is the difference between the observed value of y_{ij} and the fitted value of y_{ij}, $\hat{y}_{ij}$, so $e_{ij} = y_{ij} - \hat{y}_{ij}$. In the case of two groups, there are only two fitted values:

$$\hat{y}_{ij} = \begin{cases} \bar{y}_0 & \text{if } T_{ij} = 0, \\ \bar{y}_1 & \text{if } T_{ij} = 1. \end{cases} \qquad \text{(Equation 4.14)}$$

This means, in the two-group case, the MSE is

$$\hat{\sigma}_e^2 = \frac{\Sigma_i^{n_0} (y_{i0} - \bar{y}_1)^2 + \Sigma_i^{n_1} (y_{i1} - \bar{y}_1)^2}{n_0 + n_1 - 2}.$$

Another way to get the numerator of the sampling variance of the difference (Equation 4.12) is to use the group-specific variance of y, $\hat{\sigma}_j^2$, which is $\frac{\sum_i^{n_j}(y_{ij}-\bar{y}_j)^2}{n_j-1}$, to realize that $\sum_i(y_{ij}-\bar{y}_j)^2$ is equal to $(n-1)\hat{\sigma}_j^2$. This makes the MSE

$$\hat{\sigma}_e^2 = \frac{(n_1-1)\hat{\sigma}_1^2 + (n_0-1)\hat{\sigma}_0^2}{n_1+n_0-2},$$

which means that for bivariate models with a dichotomous predictor, the estimate of the MSE is the same as the estimate of the population variance, σ^2.

The denominator of the sampling variance (Equation 4.12) is $\sum_j\sum_i(T_{ij}-\bar{T})^2$, which is equivalent to

$$\sum_j\sum_i(T_{ij}-\bar{T})^2 = n_0\bar{T}^2 + n_1\left(1-2\bar{T}+\bar{T}^2\right) = N\bar{T}(1-\bar{T}),$$

meaning that that the estimated sampling variance of γ_1 is

$$\text{vâr}\{\hat{\gamma}_1\} = \frac{\hat{\sigma}^2}{N\bar{T}(1-\bar{T})} \qquad \text{(Equation 4.15)}$$

and its square root is the standard error

$$SE_{\gamma_1} = \frac{\hat{\sigma}}{\sqrt{N\bar{T}(1-\bar{T})}},$$

which is equivalent to Equation 3.1 since $\bar{T}=P_1$.

Slope and Intercept of Unbalanced Designs

We can see how the slope of T is the difference in group means in the unbalanced design with some algebra, using Equation 4.10, rendering

$$\hat{\gamma}_1 = \frac{\sum_i^{n_1}y_{i1} - \sum_i^{n_1}y_{i1}\bar{T} - n_1\bar{y} + n_1\bar{T}\bar{y} + \bar{T}\left(n_0\bar{y} - \sum_i^{n_0}y_{i0}\right)}{n_1 - 2n_1\bar{T} + n_0\bar{T}^2 + n_1\bar{T}^2},$$

$$\hat{\gamma}_1 = \frac{\sum_i^{n_1}y_{i1}}{N\bar{T}} - \frac{\sum_i^{n_0}y_{i0}}{N(1-\bar{T})} = \frac{\sum_i^{n_1}y_{i1}}{n_1} - \frac{\sum_i^{n_0}y_{i0}}{n_0} = \bar{y}_1 - \bar{y}_0.$$

The intercept of the model (Equation 4.11) is the mean of group 0, and we can see this with the understanding that $\gamma_1 = \bar{y}_1 - \bar{y}_0$ and $\bar{y} = \bar{T}\bar{y}_1 + (1-\bar{T})\bar{y}_0$:

$$\gamma_0 = \bar{y} - \bar{T}\gamma_1 = \bar{T}\bar{y}_1 + (1-\bar{T})\bar{y}_0 - \bar{T}(\bar{y}_1 - \bar{y}_0) = \bar{y}_0.$$

Slope and Intercept of Balanced Designs

Showing the meaning of slope and intercept is more straightforward in the balanced case. The slope can be understood when we realize that in the case of equal group sizes, the mean of T, $\bar{T}$, is one half and $(T_{ij} - \bar{T})^2$ is one quarter. This means that the denominator of Equation 4.8 is the total sample size divided by four, or half the sample, n, divided by 2,

$$\sum_j \sum_i (T_{ij} - \bar{T})^2 = \frac{N}{4} = \frac{n}{2}. \qquad \text{(Equation 4.16)}$$

The numerator of Equation 4.8 simplifies to $\frac{1}{2}\sum_i (y_{i1} - \bar{y}) - \frac{1}{2}\sum_i (y_{i0} - \bar{y})$. Thus, we can express the slope as

$$\hat{\gamma}_1 = \frac{\frac{1}{2}\sum_i (y_{i1} - \bar{y}) - \frac{1}{2}\sum_i (y_{i0} - \bar{y})}{n/2} = \frac{\sum_i (y_{i1} - \bar{y}) - \sum_i (y_{i0} - \bar{y})}{n},$$

which is equivalent to

$$\hat{\gamma}_1 = \frac{\sum_i y_{i1}}{n} - \frac{\sum_i y_{i0}}{n} = \bar{y}_1 - \bar{y}_0.$$

As for the intercept, assuming the dummy coding that makes the mean of T one half, and the fact that the overall mean, $\bar{y}$, is the average of the group means, it is relatively straightforward to see that the intercept is the mean of the group coded as $T = 0$:

$$\hat{\gamma}_0 = \frac{\bar{y}_1 + \bar{y}_0}{2} - \frac{\bar{y}_1 - \bar{y}_0}{2} = \bar{y}_0.$$

The t Test

Instead of using the normal distribution and its cumulative distribution function, analyses that also estimate the unknown parameter σ must make use of the t distribution. The implication of using this distribution in our tests is that the critical values for hypothesis testing are more extreme and change with the sample size. For example, if the sample is $N = 20$, the degrees of freedom are $v = df = 20 - 2 = 18$. Using a computer program or table to find the critical values, we find that for a two-tailed test with $\alpha = 0.05$ and $v = df = 18$, the critical value of t is 2.10, which is larger than the value for the same test using the normal distribution, which is 1.96 (see Chapter 2).

However, compared with the z test, the calculations are very much the same for the t test. The difference in means is the same. The estimated sampling variance of the difference in the means is (Equation 4.15)

$$\text{vâr}\{\bar{y}_1 - \bar{y}_0\} = \frac{\hat{\sigma}^2}{N\bar{T}(1-\bar{T})} \qquad \text{(Equation 4.17)}$$

and its square root is the standard error

$$SE_{\bar{y}_1 - \bar{y}_0} = \frac{\hat{\sigma}}{\sqrt{N\bar{T}(1-\bar{T})}},$$

which is the same expression as Equation 3.1 except now we use the *estimate* of σ. If the group sizes are the same, this expression simplifies to using only the sample size of the groups:

$$\text{vâr}\{\bar{y}_1 - \bar{y}_0\} = \hat{\sigma}^2 \frac{2}{n}. \qquad \text{(Equation 4.18)}$$

The square root of this sampling variance is the standard error of the difference in means,

$$SE_{\bar{y}_1 - \bar{y}_0} = \hat{\sigma}\sqrt{\frac{2}{n}}.$$

This makes the test of the null hypothesis (H_0) that $\bar{y}_1 - \bar{y}_0 = 0$ (if we use the absolute value of the difference to keep the test statistic positive)

$$t = \frac{|\bar{y}_1 - \bar{y}_0|}{\hat{\sigma}\sqrt{\frac{2}{n}}}$$

or, if we use the effect size (Equation 3.10) where $\delta = \frac{|\bar{y}_1 - \bar{y}_0|}{\sigma}$,

$$t = \delta\sqrt{N\bar{T}(1-\bar{T})} \qquad \text{(Equation 4.19)}$$

in the unbalanced case or

$$t = \delta\sqrt{\frac{n}{2}} \qquad \text{(Equation 4.20)}$$

in the balanced case.

Example Test With BMI Data

The BMI data table in the Appendix presents the raw data we will use in this chapter for analysis, and the summary statistics are presented in Table 4.1 for the outcome (BMI). The overall mean BMI for these observations is $\bar{y} = 33.4$ and the standard deviation for all observations is $\sigma = 5.997$. The mean for the control group is $\bar{y}_0 = 35.340$ and the mean for the treatment group is $\bar{y}_1 = 31.459$. The standard deviation for the control group is $\sigma_0 = 5.966$ and the standard deviation for the treatment group is $\sigma_1 = 5.476$.[6] Therefore, the pooled standard deviation, which is the estimate of the population standard deviation, is about $\sqrt{(5.966^2 + 5.476^2)/2} = 5.726$.[7]

The difference between the means is estimated with $\bar{y}_1 - \bar{y}_0$, which in the case of our example BMI data is $31.459 - 35.340 = -3.881$. The treatment group's mean BMI is 3.881 points lower than the mean of the control group's BMI. In our example data, the estimate of the population standard deviation (the pooled standard deviation) is $\sigma = 5.726$.[8] This makes the effect size

$$\delta = \frac{|\bar{y}_1 - \bar{y}_0|}{\sigma} = \frac{3.881}{5.726} = 0.678$$

and the test statistic (since the data are balanced with 25 observations per group)[9]

$$t = \delta\sqrt{\frac{n}{2}} = 0.678\sqrt{\frac{25}{2}} = 2.397.$$

The positive critical value for a two-tailed test is the $1 - 0.025 = 0.975$ quantile of the t distribution with 48 degrees of freedom, which is 2.011. The test statistic (2.397) exceeds 2.011, and so we can reject the null hypothesis using a two-tailed test at $\alpha = 0.05$.

The Noncentrality Parameter for Power

How could we plan for the result of this test? Perhaps we do not know exactly how the outcome scaled. That would make it difficult to plan for

---

[6] One of the major assumptions of the following statistical tests is that the variances of each group are the same. Obviously, in actual data this will never be exactly true, but there are many tests to determine whether this is a plausible assumption. Such a test was performed with the example data, and the results indicate that they are statistically the same.

[7] Note that this is smaller than the overall standard deviation (i.e., $\tilde{\sigma}^2 \neq \hat{\sigma}^2$). The overall standard deviation is equal to 5.997, whereas the pooled standard deviation is equal to 5.726. The reason is that the overall standard deviation is based on deviations from the overall mean, whereas the pooled standard deviation is based on deviations from the groups' means (cf. Equations 4.4 and 4.5).

[8] This is a little smaller than the MSE of 5.727 reported in Table 5.1 due to rounding.

[9] This does not exactly match computer output, which is 2.396, due to rounding.

Figure 4.1 Some *t* distributions with different noncentrality parameters.

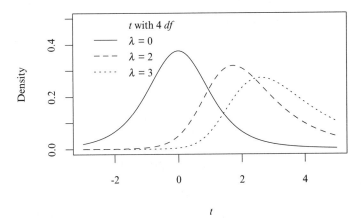

a difference in means and a pooled variance. We can make the process easier by standardizing the quantities of the test statistic into a scale-free set of parameters such as an effect size (Equation 3.10) where we replace the population standard deviation with the estimated standard deviation. By using scale-free parameters, we can make plans for the study without knowledge about the details of measurement.

Power analyses using the *t* distribution are different than power analyses that employ the normal distribution. This is because the shape of the *t* distribution is different for noncentral curves. Figure 4.1 displays a typical *t* distribution where the noncentrality parameter $\lambda = 0$, and then other curves for $\lambda = 2$ and $\lambda = 3$. Thus, the "useful relationships" employed in the case of the standard normal distribution cannot be directly applied to the case of samples (but they are still useful as we will see below).

The noncentrality parameter for the noncentral *t* distribution is the expected test statistic, assuming we knew the actual difference in means and the actual population variance. We will again call this quantity λ and express it as

$$\lambda = \underbrace{\delta}_{\text{Effect size}} \underbrace{\sqrt{N\bar{T}\left(1 - \bar{T}\right)}}_{\text{Sample size and balance}} \qquad \text{(Equation 4.21)}$$

in the unbalanced case and

$$\lambda = \underbrace{\delta}_{\text{Effect size}} \underbrace{\sqrt{\frac{n}{2}}}_{\text{Sample size}} \qquad \text{(Equation 4.22)}$$

in the balanced case. We will work with these expressions to define strategies for estimating power a priori, for finding necessary sample sizes, and for finding a minimum detectable effect.

Power Analysis for Samples Without Covariates

Finding Power A Priori

Recall from Chapter 3 that the β parameter (Type II error) for tests is the area of the alternative (noncentral) distribution before the positive critical value of the null (central) distribution for a one-tailed test and the area between the two critical values of the null (central) distribution. For a one-tailed t test, its function is

$$\beta = \underbrace{H\left[t_{(df)1-\alpha}, df, \lambda\right]}_{\text{Area before right critical value}}, \qquad \text{(Equation 4.23)}$$

and for a two-tailed t test, its function is

$$\beta = \underbrace{H\left[t_{(df)1-\frac{\alpha}{2}}, df, \lambda\right]}_{\text{Area before right critical value}} - \underbrace{H\left[t_{(df)\frac{\alpha}{2}}, df, \lambda\right]}_{\text{Area before left critical value}} \qquad \text{(Equation 4.24)}$$

where $H[a, b, c]$ is the CDF of the *noncentral t* distribution at point a with b degrees of freedom and noncentrality parameter c; $t_{(df)q}$ is the qth quantile of the central t distribution associated with the degrees of freedom (i.e., the critical value). Power, in either case, is the complement of this area:

$$\text{Power} = 1 - \beta. \qquad \text{(Equation 4.25)}$$

The H function, while it goes by different names, is available in most statistical software packages.[10]

Power is a direct function of the Type I (α) and Type II (β) error rates. For example, assuming that we expected to have an effect size of $\delta = 0.678$ and a sample size of 25 in each group, the expected test statistic would be

---

[10] For example, in base R the function is "pt," where the noncentrality parameter can be entered as one of the arguments. The function is called "nt" in Stata (there is a separate command "t" for the central distribution only), and it is called "ncdf.t" in SPSS. Each program comes with excellent installed and online help pages that explain the order of entered arguments.

Figure 4.2 Power analysis of body mass index (BMI) example result: two-tailed test with $\alpha = 0.05$ and 48 degrees of freedom so the critical value is 2.011. The expected test statistic is 2.397, so $\beta = 0.349$ (gray shaded area) and power $= 0.651$.

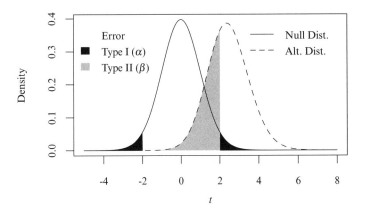

about $t = 2.397$, with $2n - 2 = 48$ degrees of freedom. That means that the noncentrality parameter (λ) is also about 2.397. Figure 4.2 visualizes this analysis, which is very similar to the process used in the normal distribution case. We find that the power of this test is 0.651.

Finding the Sample Size

Often, researchers wish to know the required number of observations necessary in order to detect an effect. Unfortunately, this is not a matter of rearranging Equation 4.23 or Equation 4.24 because the CDFs themselves include the sample size in the degrees of freedom parameter (see, e.g., Equation 4.24). Thus, we must first perform an approximation using the standard normal distribution such as Equation 3.9, which does not employ degrees of freedom. We then refine the analysis using the initial result for the degrees of freedom.

Recall from Chapter 3 that for normally distributed variables with known variances, the expected difference in means test, λ, is related to quantiles of the z distribution as (Equation 3.9)

$$\lambda \approx z_{critical} - z_\beta,$$

which leads to Equation 3.13

$$N \approx \frac{\left(z_{critical} - z_\beta\right)^2}{(P_1(1 - P_1))\delta^2}$$

in the unbalanced case and (Equation 3.14)

$$n \approx \frac{2 \left(z_{critical} - z_\beta \right)^2}{\delta^2}$$

in the balanced case.

While this is simple to do, we must remember that we are dealing with samples and the t distribution. Therefore, the standard normal approximation is only a first step, and it will tend to provide a smaller sample size than is actually required. This is because Equation 3.9 is based on the normal curve, which always has smaller quantile values than the t distribution.

One solution is to use the standard normal approximations (Equations 3.13 and 3.14) as a first step to get an initial sample size, and then use this initial sample size to find the quantiles of the appropriate t distribution based on the degrees of freedom:

$$\lambda \approx t_{(N_z-2)critical} - t_{(N_z-2)\beta}. \qquad \text{(Equation 4.26)}$$

I use the $N_z - 2$ or $2n_z - 2$ notation to remind readers that these are degrees of freedom from the standard normal approximation.

We then use this approximation in new formulas to find the necessary sample size. Thus, the formula is

$$N \approx \frac{\left(t_{(N_z-2)critical} - t_{(N_z-2)\beta} \right)^2}{\left(\bar{T} \left(1 - \bar{T} \right) \right) \delta^2} \qquad \text{(Equation 4.27)}$$

in the unbalanced case and

$$n \approx \frac{2 \left(t_{(2n_z-2)critical} - t_{(2n_z-2)\beta} \right)^2}{\delta^2} \qquad \text{(Equation 4.28)}$$

in the balanced case.

This process will still produce inexact findings, and so it is always a good idea to then use the sample size from this analysis to compute a priori power as a check. The level of approximation error decreases as the sample size becomes larger, of course, because the t distribution becomes more like the standard normal distribution after about 100 degrees of freedom.

It is always good to use a computer for the most exact computations. However, as a learning tool, quantiles for the t distribution are provided in Table 4 in the Appendix for every 1 degree of freedom from 2 to 25, and then for every 5 degrees of freedom to 100. We will use these values in the examples so that a computer will not be necessary. For example, to find the critical value for a two-tailed test with $\alpha = 0.05$ and 10 degrees of freedom, we use the row where the first column reads 10 and find the $1 - \alpha/2 = 1 - .05/2 = 1 - 0.025 = 0.975$ quantile, which is 2.228. To find the quantile associated with power of 0.8 for 10 degrees of freedom, we find the $1 - 0.8 = 0.2$ quantile on the same row, which is -0.879.

Unbalanced Case Example

Suppose we planned to perform a two-tailed test with $\alpha = 0.01$ with an effect size of $\delta = 2$, where 35% of the sample was in group 1 and thus $\overline{T} = 0.35$. We can use the procedure for the standard normal distribution (Equation 3.13) and the quantile table to find the sample size for 0.9 power ($\beta = 0.1$):

$$N_z \approx \frac{(z_{0.995} - z_{0.1})^2}{(\overline{T}(1 - \overline{T}))\,\delta^2} \approx \frac{(2.576 - (-1.282))^2}{(0.35(1 - 0.35))\,2^2} \approx 16.356,$$

or about 17 observations.

Next, we use the quantile table to find the 0.995 and 0.1 quantiles associated with $17 - 2 = 15$ degrees of freedom, which are 2.947 and -1.341, respectively. Using these values in Equation 4.27, we find the sample size to be

$$N \approx \frac{\left(t_{(N_z - 2)\text{critical}} - t_{(N_z - 2)\beta}\right)^2}{(\overline{T}(1 - \overline{T}))\,\delta^2} \approx \frac{(2.947 - (-1.341))^2}{(0.35(1 - 0.35))\,2^2} \approx 20.205,$$

or about 21 cases (about 8 in one group and 13 in the other). The actual power for this analysis is 0.930, so our approximation gave us more power than we planned.

Balanced Case Example

Now consider the same analysis as above, a two-tailed test with $\alpha = 0.01$ with an effect size of $\delta = 2$, except now we assume a balanced design. To find the sample size for each group associated with power of 0.9, we again use the procedure for the normal distribution as a first step:

$$n_z \approx \frac{2\left(z_{\text{critical}} - z_\beta\right)^2}{\delta^2} \approx \frac{2(2.576 - (-1.282))^2}{2^2} \approx 7.442,$$

or about 8 cases in each group.

Next, we use the quantile table to find the 0.995 and 0.1 quantiles associated with 14 degrees of freedom (since $2 \times 8 - 2 = 14$), which are 2.977 and -1.345, respectively. Using these values in Equation 4.28, we find the sample size to be

$$n \approx \frac{2\left(t_{(2n_z-2)critical} - t_{(2n_z-2)\beta}\right)^2}{\delta^2} \approx \frac{2\left(2.977 - (-1.345)\right)^2}{2^2} \approx 9.340,$$

or about 10 observations per group for a total of 20 observations. The actual power for this design is 0.929.

Finding the Minimum Detectable Effect

Suppose a sample size is already planned, but researchers wish to know the minimum detectable effect size (MDES) of their design for a given level of α and power $(1 - \beta$; Bloom, 1995). This quantity is the smallest effect size that a design can detect with a given power level and test, noted here as δ_m. Just as we did with standard normal distribution, we can use the quantiles of the t distribution associated with type and α of the test along with the desired Type II error (β) to find the MDES. The formula for unbalanced designs is

$$\delta_m \approx \frac{t_{(N-2)critical} - t_{(N-2)\beta}}{\sqrt{N\bar{T}(1-\bar{T})}}, \qquad \text{(Equation 4.29)}$$

and for balanced designs, it is

$$\delta_m \approx \left(t_{(2n-2)critical} - t_{(2n-2)\beta}\right)\sqrt{\frac{2}{n}}. \qquad \text{(Equation 4.30)}$$

Note that we do not need to use the standard normal curve approximations for this because we are already planning on the sample size, and thus know the degrees of freedom. I also use the degree of freedom notation of $N-2$ for unbalanced designs and $2n-2$ for balanced designs to keep the notation consistent with the type of sample size being used.

Unbalanced Case Example

Suppose we are planning a test with about 20 percent of cases in the treatment group, so $\bar{T} = 0.2$, using a one-tailed test with $\alpha = 0.05$ at power of 0.8 (so $\beta = 0.2$), and wanted to know the minimum detectable effect possible with 82 cases. Using the quantile table, we know that there are $N - 2 = 80$ degrees of freedom, and so $t_{(2n-2)critical} = t_{(80)1-.05} =$

$t_{(80)0.95} = 1.664$ and $t_{(2n-2)\beta} = t_{(80)0.2} = -0.846$. Thus, the computation for the MDES is

$$\delta_m \approx \frac{t_{(N-2)critical} - t_{(N-2)\beta}}{\sqrt{N\bar{T}(1-\bar{T})}} \approx \frac{1.664 - (-0.846)}{\sqrt{82 \times 0.2(1-0.2)}} \approx 0.693.$$

This indicates that this design has the ability to detect an effect size of 0.693 standard deviations.

Balanced Case Example

Suppose we were planning a one-tailed test with $\alpha = 0.05$ at power of 0.8 (so $\beta = 0.2$) and wanted to know the minimum effect possible with 82 cases. This time, the sample is balanced with 41 cases in each group. Again, using the quantile table, we know that there are $2n - 2 = 80$ degrees of freedom, and so $t_{(2n-2)critical} = t_{(80)1-.05} = t_{(80)0.95} = 1.664$ and $t_{(2n-2)\beta} = t_{(80)0.2} = -0.846$. Thus, the computation for the MDES is

$$\delta_m \approx \left(t_{(2n-2)critical} - t_{(2n-2)\beta}\right) \sqrt{\frac{2}{n}} \approx (1.664 - (-0.846)) \sqrt{\frac{2}{41}} \approx 0.554.$$

This indicates that this design has the ability to detect an effect size of 0.554 standard deviations.

Influences on Power

Through the examples, it is apparent that many parameters influence power. If we hold constant the type of test (it is usually two-tailed), the level of significance (it is usually $\alpha = 0.05$), and balance (if we are planning experiments, we typically assume a balanced design to maximize power), we can draw graphs to better understand the relationship between effect size, sample size, and power.

The effect size has a large influence on power. As you can see in Figure 4.3, which assumes a two-tailed test with $\alpha = 0.05$, increasing the sample size (n) always increases power. However, the curves are flatter for lower effect sizes. For example, even with 40 observations in each group, power is low for an effect size of 0.5. Conversely, with a larger effect size such as 1, adequate power is reached with relatively few observations per group.

Inversely, power and sample size also have an effect on the minimum detectable effect size. As you can see in Figure 4.4, which assumes a two-tailed test with $\alpha = 0.05$, increasing the sample size (n) always lowers the

52

Figure 4.3 Power curves by *n* and effect size (δ) for a two-tailed *t* test with a sample of 2*n* and $\alpha = 0.05$.

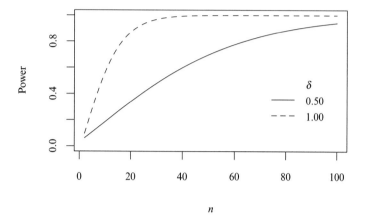

Figure 4.4 Minimum detectable effect size (MDES) by *n* and power for a two-tailed *t* test with a sample of 2*n* and $\alpha = 0.05$.

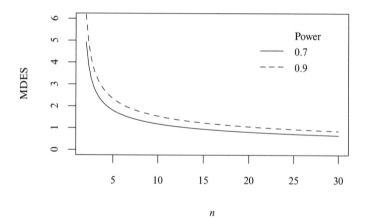

possible effect size that can be detected. Looking at the two lines, we see that the lower the power, the lower the detectable effect size.

Summary

This chapter considered how to analyze the power for testing the difference between two group means in a simple random sample. We started with the basic methods to perform this test using regression. We then defined the parameters for a power analysis and explored different computations under balanced and non-balanced designs. As we find that designs that are not balanced have less power, we will work with balanced designs moving forward.

In the next chapter, covariates are introduced. We again use regression to understand the effect that covariates have on the test of the difference between group means.

CHAPTER 5. USING COVARIATES WHEN TESTING THE DIFFERENCE IN SAMPLE GROUP MEANS FOR BALANCED DESIGNS

In this chapter, I expand the analysis of group differences in means to include the use of covariates. The effect that a covariate has on power can be both beneficial and harmful, depending on the correlation between the treatment indicator variable and the covariate. The parameters for power are outlined for both the case in which there is correlation between treatment indicator and the covariate, and the case in which the treatment indicator and the covariate are uncorrelated due to randomization of the treatment condition. For brevity, we consider only balanced designs.

While this chapter goes into the test and its parameters in detail, a conceptual treatment of the use of covariates in the context of two-group designs may be helpful. Remember that statistical power is based on the statistical test, which in turn is based on the sampling variances of the effects. Anything that makes the sampling variances smaller will lead to a more powerful test.

When it is possible to find variables that are correlated with outcomes but are not correlated with other predictors, these variables can be used in regression models to reduce the residual variance. Reducing this variance improves power. Recall the formula for the sampling variance of a regression coefficient (Equation 4.12):

$$\text{vâr}\{\hat{\gamma}_1\} = \frac{\hat{\sigma}_e^2}{\sum_j \sum_i \left(T_{ij} - \overline{T}\right)^2}.$$

This expression means that the sampling variance (which we use to get the standard error) of a regression slope is based on the variance of the residuals, $\hat{\sigma}_e^2$. The smaller the variance in the residuals, the smaller the sampling variance of the coefficient. The variance of residuals always gets smaller when models include more covariates. Without correlations between predictors, this means that we can always make sampling variances get smaller with covariates.

To visualize how regression leads to smaller variances, consider the two histograms in Figure 5.1. On the left is a histogram of some arbitrary variable. Next, on the right, is a histogram of residuals from a model where this variable is regressed on another arbitrary predictor. As you

Figure 5.1 Histograms and standard deviations of a random outcome and its residuals from a regression model. Note that the residuals have a smaller standard deviation.

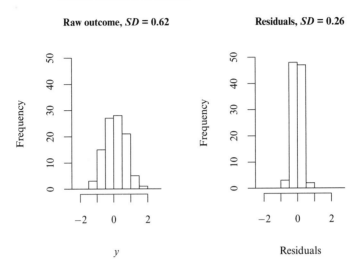

In this text, **Raw outcome, *SD* = 0.62** and **Residuals, *SD* = 0.26**

can see, the variance (spread) of the residuals is smaller on the right than the raw outcome on the left.

In the next section, we will see how sometimes this does not improve power due to correlations between predictors.

Example Analysis

We begin with a return to our BMI data. As you will recall from Chapter 4, we found a significant difference in the BMI scores when we compared those who ate breakfast with those who did not eat breakfast. This result is replicated in Table 5.1 under the "Treatment Only" regression model. In the next regression model, we add a pretest that was measured at the beginning of the study. The results are reported in the "Treatment With Pretest" model. Here, the effect is much smaller and no longer significant.

The reason for the small and insignificant effect is revealed in the summary statistics of the pretest, presented in Table 5.2. Here, we see that the average BMI for those assigned to control is higher than for those who were assigned to treatment. This means that there is a correlation between

Table 5.1   Models Predicting BMI

	Treatment Only	Treatment With Pretest
Intercept	35.340***	−0.617
	(1.145)	(0.992)
Breakfast vs. no breakfast	−3.881*	−0.084
	(1.620)	(0.315)
Pretest		1.017***
		(0.027)
R^2	0.107	0.970
N	50	50
Root MSE	5.727	1.052

Note. BMI = body mass index; MSE = mean squared error. Standard errors in parentheses. *$p < 0.05$. **$p < 0.01$. ***$p < 0.001$.

Table 5.2   Summary Statistics of Body Mass Index Scores Before Treatment

	N	Mean	Standard Deviation
No breakfast (control)	25.000	35.360	5.586
Breakfast (treatment)	25.000	31.626	5.489
Total sample	50.000	33.493	5.796

the treatment indicator and the covariate, which causes multicolinearity and reduces power. When randomization works, and this correlation is not present, adding a covariate improves power considerably.

Tests Employing a Covariate (ANCOVA) With Balanced Samples

In many cases, researchers will wish to employ a covariate in their analysis of group differences in means. This procedure is called analysis of covariance, or ANCOVA. ANCOVA is employed primarily to increase precision in randomized experiments, but it can also be employed in quasi-experimental settings to remove spurious effects, or simply in observational studies that are interested in the effects of factors, but wish to add control variables (Wildt & Ahtola, 1978).

However, employing covariates can be a risky proposition for power because any correlations (even by happenstance) between the treatment indicator and the covariate will alter both the estimate of the difference in means and its sampling variance. Generally, this leads to a smaller test statistic. However, if it is possible to ensure that the correlation between the treatment indicator and the covariate is near zero, the gains in power can be large if the correlation between the covariate and outcome is also large.

Defining the Test

We start with the data-generating process, adding a new term to Equation 4.1:

$$y_{ij} = \mu + \tau'_j + \phi_1\left(x_{ij} - \bar{x}\right) + e_{ij}, \qquad \text{(Equation 5.1)}$$

where τ'_j is the adjusted treatment effect and ϕ_1 is the within-group slope for x, the covariate, predicting the outcome, y.[1] An estimate of ϕ_1 is possible using a within-effects regression model (also known to econometritions as a fixed effects model):

$$y_{ij} - \bar{y}_j + \bar{y} = \hat{\phi}_0 + \hat{\phi}_1\left(x_{ij} - \bar{x}_j + \bar{x}\right) + r_{ij}. \qquad \text{(Equation 5.2)}$$

From Equation 5.1 it can be shown that the outcome can be adjusted for its relationship with the covariate to return to a form similar to Equation 4.1:

$$y_{ij} - \phi_1\left(x_{ij} - \bar{x}\right) = \mu + \tau'_j + e_{ij}. \qquad \text{(Equation 5.3)}$$

This means that the adjusted treatment effect, τ'_j, is related to the unadjusted treatment effect through ϕ_1 and the difference between the means of the covariate by treatment and control groups (Porter & Raudenbush, 1987):

$$\tau'_j = \tau_j - \phi_1\left(\bar{x}_1 - \bar{x}_0\right). \qquad \text{(Equation 5.4)}$$

When there is no correlation between T and the covariate x, the means of x for each group are the same. In such cases, $\tau'_j = \tau_j$.

ANCOVA From Regression

In this section, we consider the test statistic from a multiple regression framework, which allows us to break down the noncentrality parameter into effect sizes and correlations.

---

[1] That is, the population slope from the regression of y on x without introducing a treatment.

Before, we considered a linear regression model to estimate the difference in means (Equation 4.7). Here, we add the effect of the mean-centered covariate to the regression model:

$$y_{ij} = \gamma_0 + \gamma_1 T_{ij} + \phi_1 \left(x_{ij} - \bar{x}\right) + e_{ij}. \qquad \text{(Equation 5.5)}$$

The estimation of this model with least squares simultaneously estimates the adjusted difference in means and the within-group effect of x on y. This has an effect on the estimate of the difference in means. The adjusted difference in means is no longer $\hat{\gamma}_1 = \bar{y}_1 - \bar{y}_0$, but is instead

$$\hat{\gamma}_1 = (\bar{y}_1 - \bar{y}_0) - \phi_1 (\bar{x}_1 - \bar{x}_0). \qquad \text{(Equation 5.6)}$$

Next, the sampling variance of the adjusted difference in means is no longer $\sigma^2 \frac{2}{n}$. This is because the estimate of σ^2 is no longer equivalent to the estimate of the mean squared error, σ_e^2, due to the introduction of the covariate. The estimate of the MSE for the regression is now (Kirk, 1995)

$$\hat{\sigma}_e^2 = \hat{\sigma}^2 \left(1 - \hat{\rho}_{yx,w}^2\right) \frac{pn - p}{pn - p - 1} \qquad \text{(Equation 5.7)}$$

where $\rho_{yx,w}$ is the within-group correlation between the outcome, y, and covariate, x, estimated as[2]

$$\hat{\rho}_{yx,w} = \frac{\sum_j \sum_i \left(y_{ij} - \bar{y}_j\right) \left(x_{ij} - \bar{x}_j\right)}{\sqrt{\sum_j \sum_i \left(y_{ij} - \bar{y}_j\right)^2 \sum_j \sum_i \left(x_{ij} - \bar{x}_j\right)^2}} \qquad \text{(Equation 5.8)}$$

and $\frac{pn-p}{pn-p-1}$ is an adjustment to the pooled standard deviation that is necessary because the covariate removes a degree of freedom. The best way to think about $\rho_{yx,w}$ is that it is the estimate of the population correlation between y and x assuming no interventions or group effects. The sample correlation between y and x is not a good estimate because of the induced treatment effects on y.

It is well known that the the sampling variance of any slope, say for variable q, in a multivariate regression is (see, e.g., Fox, 2015)

$$\text{var}\left\{\gamma_q\right\} = \frac{\sigma_e^2}{\sum \left(x_q - \bar{x}_q\right)^2} \frac{1}{1 - \rho_{qx}^2} \qquad \text{(Equation 5.9)}$$

---

[2] Note that, like the pooled variance, the estimate of the population correlation between the outcome and the covariate is based on deviations from the group means. In other words, this is the correlation between y and x if no treatment was introduced.

where σ_e^2 is the mean squared error of the regression, $\sum (x_q - \bar{x}_q)^2$ is the sum of squares for the predictor, and $\frac{1}{1-\rho_{qx}^2}$ is a variance inflation factor (VIF) due to the multiple correlation between the predictor q and all other covariates x. The VIF is a factor that increases the sampling variance of a regression effect proportionate to the relationship between the predictor and other covariates. The stronger the relationship among predictors, the larger this factor inflates the sampling variances. In the case of the predictor T, we know $\sum_j \sum_i (T_{ij} - \bar{T})^2$ is equal to $\frac{n}{2}$ in balanced samples (Equation 4.16). In our case, where the coefficient of interest is γ_1, the effect of T on y, with a covariate x and equal group sizes, this expression becomes

$$\text{var}\{\gamma_1\} = \sigma_e^2 \frac{2}{n} \frac{1}{1 - \rho_{Tx}^2}. \qquad \text{(Equation 5.10)}$$

Thus, if we replace σ_e^2 with Equation 5.7, this sampling variance is now (assuming two groups of equal sample sizes)

$$\hat{\text{var}}\{\hat{\gamma}_1\} = \hat{\sigma}^2 \left(1 - \hat{\rho}_{yx,w}^2\right) \frac{2n-2}{2n-3} \frac{2}{n} \frac{1}{1 - \hat{\rho}_{Tx}^2} \qquad \text{(Equation 5.11)}$$

where $\hat{\sigma}^2$ is the estimated population variance from Equation 4.6, which is reduced by a factor of $1 - \hat{\rho}_{yx,w}^2$. Finally, the factor $\frac{1}{1-\hat{\rho}_{Tx}^2}$ is the VIF due to the overall squared correlation between the treatment indicator and the covariate (or the proportion of variance in T explained by x).

Example Analysis With BMI Data

The first step of the process to test the coefficient γ_1 is to estimate the within-group slope of y regressed on x, the within-group correlation between y and x, and the correlation between the treatment indicator and the covariate. In the example BMI data, these estimates are $\hat{\phi}_1 = 1.017$, $\hat{\rho}_{yx,w} = 0.9833$, and $\hat{\rho}_{Tx} = -0.325$, respectively.

The adjusted difference in these data can be calculated from the values in Table 4.1, Table 5.2, and $\hat{\phi}_1$ using Equation 5.6:

$$\hat{\gamma}_1 = -3.881 - 1.017 \times (31.626 - 35.360) = -0.084.$$

To calculate the estimated sampling variance of this difference, we use Equation 5.11,

$$\hat{\text{var}}\{\hat{\gamma}_1\} = 5.727^2 (1 - 0.9833^2) \frac{50-2}{50-3} \frac{2}{25} \frac{1}{1 - (-0.325)^2} = 0.099,$$

the square root of which is the standard error of 0.315 reported in Table 5.1.

Finally, the t test is calculated as

$$t = \frac{|-0.084|}{0.315} = 0.267,$$

which has a two-tailed p value of 0.605 with 47 degrees of freedom, which is nowhere near a convention for statistical significance. The following discussion will serve as a guide to what happened.

Defining the Parameters for Power

The regression framework lends itself to putting together a noncentrality parameter since we can use the slope to construct an effect size analogous to the case without a covariate. Using Equations 5.6 and 5.11 we can state the noncentrality parameter as

$$\lambda = \frac{(\mu_1 - \mu_0) - \phi_1 (\bar{x}_1 - \bar{x}_0)}{\sqrt{\sigma^2 \left(1 - \rho_{yx,w}^2\right) \frac{2n-2}{2n-3} \frac{2}{n} \frac{1}{1-\rho_{Tx}^2}}},$$

which can be reorganized as the product of an adjusted effect size and the square roots of the the reciprocal of the covariate effect, the sample size, and the effect of multicolinearity:

$$\lambda = \underbrace{\frac{(\mu_1 - \mu_0) - \phi_1 (\bar{x}_1 - \bar{x}_0)}{\sigma}}_{\text{Adjusted effect size}} \underbrace{\sqrt{\frac{1}{\left(1 - \rho_{yx,w}^2\right)}}}_{\text{Covariate effect}} \underbrace{\sqrt{\frac{n}{2} \frac{2n-3}{2n-2}}}_{\text{Sample size}} \underbrace{\sqrt{1 - \rho_{Tx}^2}}_{\text{Multicolinearity}}.$$

(Equation 5.12)

Power analysis without covariates (Equation 4.22) has only two parameters, the effect size and sample size. Conversely, power analyses for the case of a correlated covariate requires a substantial amount of prior information as seen in Equation 5.12, including the effectiveness of the covariate in explaining variation in the outcome ($\rho_{yx,w}^2$) and some idea of the correlation that can be expected between the group indicator and the covariate (ρ_{Tx}). An expectation of the adjusted effect size is

$$\delta_a = \frac{(\mu_1 - \mu_0) - \phi_1 (\bar{x}_1 - \bar{x}_0)}{\sigma}.$$

(Equation 5.13)

The sample size can be simplified as well. The factor $\frac{2n-3}{2n-2}$ is always less than 1, but it gets close to 1 as samples reach $n = 30$. However, the square root of this expression never dips below 0.95 for samples of $n = 6$ or greater. Thus, unless the sample is extremely small, the effect of this factor on the noncentrality parameter is minimal. Knowing this and an effect size, Equation 5.12 can be approximated with the simplified formula

$$\lambda \approx \delta_a \sqrt{\frac{n}{2}} \sqrt{\frac{1}{(1 - \rho_{yx,w}^2)}} \sqrt{1 - \rho_{Tx}^2}. \qquad \text{(Equation 5.14)}$$

Power Analysis With a Covariate Correlated With the Treatment Indicator

Power analysis when the covariate is correlated with the treatment indicator is difficult. The nature of the correlation between the treatment indicator and the covariate is even harder to anticipate than the effect size.

Anticipating the Correlation Between Treatment and the Covariate

One possibility to make a guess about the correlation between the treatment variable and the covariate tractable is to posit an effect size for the covariate. It is well known that the correlation between two variables is a product of the slope and ratio of standard deviations,

$$\rho_{ab} = \gamma_{ab} \frac{\sigma_b}{\sigma_a}.$$

Since the correlation between variables a and b is the same as the correlation between b and a, and since (in the case of balanced group sizes) the slope of T predicting x is $\bar{x}_1 - \bar{x}_0$, we can express ρ_{Tx} as

$$\rho_{Tx} = (\bar{x}_1 - \bar{x}_0) \frac{\sigma_T}{\sigma_x}.$$

If T is a dichotomous indicator of balanced groups, we can express this as

$$\rho_{Tx} = \frac{(\bar{x}_1 - \bar{x}_0)}{\sigma_x} \sqrt{\frac{pn}{4(pn-1)}} = \delta_{\bar{x}_1 - \bar{x}_0} \sqrt{\frac{pn}{4(pn-1)}},$$

where $\delta_{\bar{x}_1 - \bar{x}_0}$ is an effect size measuring the standardized difference in the covariate between the treatment and control groups, which means that we can approximate the term ρ_{Tx}^2 in large balanced samples as

$$\rho_{Tx}^2 \approx \frac{\delta_{\bar{x}_1 - \bar{x}_0}^2}{4}. \qquad \text{(Equation 5.15)}$$

For example, looking at Table 5.2, the covariate's difference between treatment and control was $31.626 - 35.360 = -3.734$, and that divided by the overall standard deviation of x is $\delta_{\bar{x}_1 - \bar{x}_0} = -3.734/5.796 = -0.644$. If we square this effect size we get 0.415, and that divided by 4 is 0.104. This is the approximation of ρ_{Tx}^2.[3] It is also worth noting that this example means that only 10% of the variance in the covariate is explained by the treatment indicator, which is not a very large effect. Yet it was enough to derail the analysis and remove the significance of the effect.

Finding Power A Priori

The adjusted effect size (Equation 5.13) from the above analysis using the example data is about (ignoring the direction of the effect) $|\hat{\delta}_a| = 0.084/5.727 = 0.0147$. Suppose we also expected to have a covariate that had a within-group correlation with the outcome of $\rho_{yx,w} = 0.9833$ and that had an overall correlation with the treatment indicator of $\rho_{Tx} = -0.325$. Given the low effect size and the nonzero correlation between treatment and the covariate, the result was a small t-statistic of $|t| = 0.267$. This is also the value of λ,

$$\lambda = \frac{(\mu_1 - \mu_0) - \phi_1(\bar{x}_1 - \bar{x}_0)}{\sigma}\sqrt{\frac{1}{(1 - \rho_{yx,w}^2)}}\sqrt{\frac{n}{2}\frac{2n - 3}{2n - 2}}\sqrt{1 - \rho_{Tx}^2},$$

$$\lambda = 0.0147\sqrt{\frac{1}{(1 - 0.9833^2)}}\sqrt{\frac{25}{2}\frac{50 - 3}{50 - 2}}\sqrt{1 - (-0.325)^2} = 0.267,$$

and using Equation 4.24, we find that the power is about 0.058 (see Figure 5.2).

Again, when a correlated covariate is employed, the power analysis must work with several parameters in addition to the (adjusted) effect size and sample size, including the correlation between the outcome and covariate ($\rho_{yx,w}$) and the correlation between the treatment indicator and the covariate (ρ_{Tx}). The magnitudes of each correlation have opposite effects on power.

As we saw in Figure 4.3, power increases with the sample size. However, as the correlation between the treatment indicator and the covariate

---

[3] The square root of this quantity is -0.322, which is very close to what we observed, $\rho_{Tx} = -0.325$.

Figure 5.2 Power analysis of body mass index treatment result when covariate is employed. $\delta_a = 0.0147$, $\lambda = 0.267$, $\beta = 0.942$, power = 0.058.

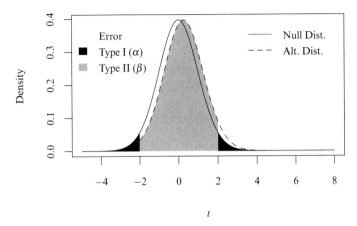

increases, the power for any given sample size decreases. Unlike the effect of the correlation between the treatment indicator and the covariate, as the correlation between the outcome and the covariate increases, the power for any given sample size increases.

Figure 5.3 presents the effect of the correlation between the treatment indicator and the covariate on power, assuming $\delta_a = 0.50$, $\rho_{yx,w} = 0.75$,

Figure 5.3 Power curves by n and ρ_{Tx} for a two-tailed t test with $\alpha = 0.05$, $\rho_{yx,w} = 0.75$, and $\delta_a = 0.5$.

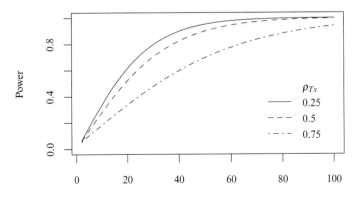

64

and a two-tailed test with $\alpha = 0.05$. As we saw in Figure 4.3, power increases with the sample size. However, as the correlation between the treatment indicator and the covariate increases, the power for any given sample size decreases. For example, when n is about 40, the correlation between the treatment indicator and the covariate is about 0.25, the adjusted effect size is about 0.5, and the correlation between the covariate and outcome is about 0.75, the power of the test is about 0.9. Power is affected somewhat when this correlation between the treatment indicator and the covariate increases to 0.5. The impact on power is large if the correlation between the treatment indicator and the covariate is 0.75: It reduces power to about 0.6.

Figure 5.4 presents the effect of the correlation between the outcome and the covariate on power assuming $\delta_a = 0.50$, $\rho_{Tx} = 0.50$, and a two-tailed test with $\alpha = 0.05$. Unlike the effect of the correlation between the treatment indicator and the covariate, as the correlation between the outcome and the covariate increases, the power for any given sample size increases. For example, when n is about 40, the correlation between the outcome and the covariate is about 0.25, the adjusted effect size is about 0.5, and the correlation between the treatment indicator and covariate is about 0.5, the power of the test is about 0.5. Power is affected somewhat when the correlation between the outcome and the covariate increases to 0.5. However, the impact on power is positive if this correlation between the outcome and the covariate is 0.75, increasing power to over 0.8.

Figure 5.4   Power curves by n and $\rho_{yx,w}$ for a two-tailed t test with $\alpha = 0.05$, $\rho_{Tx} = 0.5$, and $\delta_a = 0.5$.

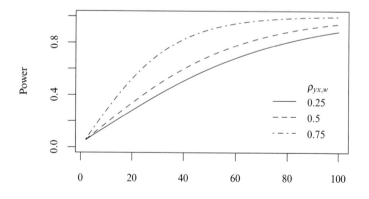

Figure 5.5 Power curves by $\rho_{yx,w}$ and ρ_{Tx} for a two-tailed t test with $\alpha = 0.05$ and $n = 40$, $\delta_a = 0.5$.

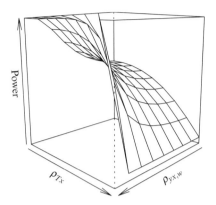

Figure 5.5 presents the joint effects of the correlation between the treatment indicator and the covariate and the correlation between the outcome and the covariate assuming $n = 40$, $\delta_a = 0.50$, and a two-tailed test with $\alpha = 0.05$. Following the curves lining up with the correlation between the treatment indicator and the covariate (ρ_{Tx}), we see that power quickly decreases with this correlation in proportion to the power when this correlation is 0. Following the curves lining up with the correlation between the outcome and the covariate ($\rho_{yx,w}$), power increases quickly after about 0.4. From this, we can conclude that it is best to avoid correlations between the treatment indicator and the covariate. However, it is a good idea to use a covariate if the correlation between it and the outcome is at least moderate (greater than 0.4).

Finding the Sample Size

Assuming prior knowledge provides insight on the correlations between the covariate, outcome, and treatment indicator, we can rearrange Equation 5.14 using Equation 3.9 to provide an approximate sample size formula,

$$n_z \approx \frac{2\left(z_{critical} - z_\beta\right)^2}{\delta_a^2} \frac{1 - \rho_{yx,w}^2}{1 - \rho_{Tx}^2}, \qquad \text{(Equation 5.16)}$$

that provides a starting sample size for degrees of freedom that can be used in

$$n \approx \frac{2\left(t_{(2n_z-2)critical} - t_{(2n_z-2)\beta}\right)^2}{\delta_a^2} \frac{1 - \rho_{yx,w}^2}{1 - \rho_{Tx}^2}. \qquad \text{(Equation 5.17)}$$

This expression is informative because it shows not only that as the effect size (δ_a) increases, the sample size will decrease, but also that the effectiveness of the effect size to reduce the sample size can be scaled back when there is correlation between the treatment indicator and the covariate (ρ_{Tx}). Finally, the sample size can again be decreased as the correlation between y and x ($\rho_{yx,w}$) increases.

Example

For example, suppose that we wish to find the sample size for each group in a balanced design to detect an adjusted effect size of $\delta_a = 0.7$, where the correlation between the outcome and covariate was about $\rho_{yx,w} = 0.8$ but the correlation between the treatment groups and covariate was smaller at $\rho_{Tx} = 0.1$. We first determine an approximate sample size using the quantiles of the standard normal distribution. If we assume a two-tailed test at $\alpha = 0.05$ and wish to have 0.8 power, the useful quantiles of the standard normal distribution are $z_{0.975} = 1.96$ and $z_{0.2} = -0.842$, thus the sample size is approximately

$$n_z \approx \frac{2(1.96 - (-0.842))^2}{0.7^2} \frac{1 - 0.8^2}{1 - 0.1^2} \approx 11.653$$

or about 12 cases per group. We then use this with the appropriate t distribution quantiles with $2 \times 12 - 3 = 21$ degrees of freedom, where $t_{(21)0.975} = 2.08$ and $t_{(21)0.2} = -0.859$, and get

$$n \approx \frac{2(2.08 - (-0.859))^2}{0.7^2} \frac{1 - 0.8^2}{1 - 0.1^2} \approx 12.820,$$

or about 13 cases per group.

Finding the Minimum Detectable Effect

We can again rearrange Equation 5.14 to isolate the minimum detectable adjusted effect size

$$\delta_{a,m} \approx \left(t_{(2n-2)critical} - t_{(2n-2)\beta}\right) \sqrt{\frac{2}{n}} \sqrt{\frac{1 - \rho_{yx,w}^2}{1 - \rho_{Tx}^2}}. \qquad \text{(Equation 5.18)}$$

This expression is informative because it illustrates the relative effect of the correlations between the outcome and covariate and the treatment indicator and the covariate. The effect of the correlation between the outcome and covariate is doubled, while the effect of the correlation between the treatment indicator and the covariate is multiplied by the group sample size. Thus, assuming equal correlations, the effect of the correlation between the treatment indicator and the covariate is far stronger, as the sample size increases.

Example

We again use the quantiles of the t distribution to find the minimum detectable effect with Equation 5.18. Suppose that for a two-tailed test at $\alpha = 0.05$ with power of 0.8 (so $\beta = 0.2$) we planned a balanced study with $n = 24$ cases per group, so we have $2n - 3 = 45$ degrees of freedom. We also planned on the correlation between the outcome and covariate to be about $\rho_{yx,w} = 0.8$ and the correlation between the treatment groups and covariate to be smaller at $\rho_{Tx} = 0.1$. The MDES for this design, where $t_{(45)0.975} = 2.014$ and $t_{(45)0.2} = -0.85$, is

$$\delta_{a,m} \approx (2.014 - (-0.85)) \sqrt{\frac{2}{24}} \sqrt{\frac{1 - 0.8^2}{1 - 0.1^2}} \approx 0.499,$$

or about a 0.499 adjusted effect size.

Power Analysis With a Covariate Uncorrelated to the Treatment Indicator

Power analyses for the case of a covariate uncorrelated with the treatment indicator require less prior information, as seen in Equation 5.14. It includes only the effectiveness of the covariate in explaining variation $(\rho_{yx,w}^2)$. Moreover, because the correlation between the treatment effect and the covariate is assumed to be 0, the effect size is no longer adjusted and returns to[4]

$$\delta = \frac{\mu_1 - \mu_0}{\sigma}.$$

---

[4] This is not the only way to do this power analysis. Lipsey (1990) suggested using $\frac{\bar{y}_1 - \bar{y}_0}{\sqrt{\sigma^2(1-\rho_{\bar{y}x,w}^2)}}$ to calculate the larger effect size that is obtained from using a covariate uncorrelated with the treatment indicator. In the case of an uncorrelated covariate, this allows researchers to employ standard power tables.

Also, the variance inflation factor is removed so Equation 5.12 can be approximated with the simplified formula

$$\lambda \approx \underbrace{\delta}_{\text{Effect size}} \underbrace{\sqrt{\frac{n}{2}}}_{\text{Sample size}} \underbrace{\sqrt{\frac{1}{\left(1 - \rho_{yx,w}^2\right)}}}_{\text{Covariate effect}} . \qquad \text{(Equation 5.19)}$$

In summary, power analysis in the case of a covariate uncorrelated with the treatment indicator involves only the effect size, sample size, and the effectiveness of the covariate in explaining variation in the outcome.

Finding Power A Priori

Finding power in this situation is no different than in other situations. Essentially, once the noncentrality parameter is computed, we use the noncentral t distribution along with the appropriate degrees of freedom to determine the area associated with the Type II error (β). Once that is determined, power is simply the complement, $1 - \beta$.

Example

Suppose in our example data that there was no correlation between the treatment assignment and the covariate. Recall the effect size from the unadjusted analysis using the example data is about $\delta = 0.678$. Also suppose we expected to have a covariate that had a correlation with the outcome of $\rho_{yx,w} = 0.9833$ but no correlation with the treatment indicator so $\rho_{Tx} = 0.0$. The noncentrality parameter for the original analysis was 2.397, which was associated with power of 0.651 with 48 degrees of freedom. If we had a covariate that was highly correlated with the outcome such that $\rho_{yx,w} = 0.9833$, the new noncentrality parameter would be

$$\lambda \approx \delta \sqrt{\frac{n}{2}} \sqrt{\frac{1}{\left(1 - \rho_{yx,w}^2\right)}} \approx 0.678 \sqrt{\frac{25}{2}} \sqrt{\frac{1}{1 - 0.9833^2}} \approx 13.171.$$

This would be our t statistic, which is associated with a power of nearly 1.

Finding the Sample Size

Assuming prior knowledge provides insight about the correlation between the covariate and the outcome, we can rearrange Equation 5.19 using Equation 3.9 to provide an approximate sample size formula

$$n_z \approx \frac{2\left(z_{critical} - z_\beta\right)^2}{\delta^2} \left(1 - \rho_{yx,w}^2\right). \qquad \text{(Equation 5.20)}$$

We can then use the results of this formula to find a better approximate sample size:

$$n \approx \frac{2\left(t_{(2n_z-2)critical} - t_{(2n_z-2)\beta}\right)^2}{\delta^2}\left(1 - \rho_{yx,w}^2\right). \qquad \text{(Equation 5.21)}$$

Example

Suppose we expected an effect size of $\delta = 0.75$ and believed we could use a covariate that was uncorrelated with the treatment variable that would be correlated with the outcome at $\rho_{yx,w} = 0.8$. If we wanted to test our hypothesis with an $\alpha = 0.01$ two-tailed test and wanted to achieve power of 0.9 (so $\beta = 0.1$), we would first approximate the sample size using Equation 5.20 and $z_{0.995}$ and $z_{0.1}$,

$$n_z \approx \frac{2\left(z_{critical} - z_\beta\right)^2}{\delta^2}\left(1 - \rho_{yx,w}^2\right),$$

$$n_z \approx \frac{2\left(2.576 - (-1.282)\right)^2}{0.75^2}\left(1 - 0.8^2\right) = 19.052,$$

or about 20 cases per group.

We then use this value with our quantile table, taking the row of 35 degrees of freedom[5] for t quantile values of $t_{(35)0.995}$ and $t_{(35)0.1}$,

$$n \approx \frac{2\left(t_{(35)0.995} - t_{(35)0.1}\right)^2}{\delta^2}\left(1 - \rho_{yx,w}^2\right),$$

$$n \approx \frac{2\left(2.724 - (-1.306)\right)^2}{0.75^2}\left(1 - 0.8^2\right) \approx 20.788,$$

or about 21 cases per group.

Finding the Minimum Detectable Effect

We can again rearrange Equation 5.19 to isolate the MDES:

$$\delta_m \approx \left(t_{(2n-3)critical} - t_{(2n-3)\beta}\right)\sqrt{\frac{2}{n}}\sqrt{1 - \rho_{yx,w}^2}. \qquad \text{(Equation 5.22)}$$

---

[5] We should use the degrees of freedom of $2 \times 20 - 3 = 37$ and a computer, since this entry is not in the quantile table, but we can round down for this example.

Example

Suppose we knew that we would have a sample size of $n = 12$ in each group, and believed we could use a covariate that was uncorrelated with the treatment variable that would be correlated with the outcome at $\rho_{yx,w} = 0.8$. If we wanted to test our hypothesis with an $\alpha = 0.01$ two-tailed test and wanted to achieve power of 0.9 (so $\beta = 0.1$), we would use $t_{(22)0.995}$ and $t_{(22)0.1}$. The smallest possible effect that we could detect would be, using Equation 5.22,

$$\delta_m \approx (2.831 - (-1.323)) \sqrt{\frac{2}{12}} \sqrt{1 - 0.8^2} \approx 1.018.$$

This means that we would need an effect of more than a standard deviation in order to meet the criteria for a successful test.

Summary

The previous chapter dealt with testing the difference of two group means using a simple t test. This chapter explored the impact that a covariate would have on such an analysis by examining the ANCOVA model through regression. We found that covariates can be helpful because they reduce the error variance of the regression model, and this in turn reduces the sampling variance of the regression effect. However, if there is correlation between the covariate and the treatment indicator, another factor plays into the sampling variance: multicollinearity. The VIF, in this context, is based on the correlation between the treatment indicator and the covariate. As this correlation increases, the VIF increases the sampling variance of the regression effects, resulting in less statistical power. In our example data, we witnessed such a correlation, and as a result the benefits of having a covariate were outweighed by the effects of multicollinearity.

The next chapter is again focused on group mean differences. However, we now complicate matters with a more complex, multilevel, sampling design. This multilevel sampling design also can include covariates at different levels.

CHAPTER 6. MULTILEVEL MODELS I: TESTING THE DIFFERENCE IN GROUP MEANS IN TWO-LEVEL CLUSTER RANDOMIZED TRIALS

In this chapter, the topic of cluster, or multilevel, designs is introduced. Multilevel designs are common in experimental research (for a good review of reasons, see Bloom, 2005). Multilevel designs are especially common in education (O'Connell & McCoach, 2008) and health (Donner & Klar, 2000; Murray, 1998). Although multilevel modeling software such as "HLM" (Raudenbush, Bryk, & Congdon, 2004) is typically used to fit such models, we explore these tests using an ANOVA framework to better understand the mechanics and find parameters for power.[1] The focus in this chapter will be on two-level designs, but the reader is encouraged to seek out further reading on the subject for designs with more than two levels (see, e.g., Hedges & Rhoads, 2010). The general approach for this chapter is to give more detail for simple analyses, and then provide only the crucial detail as the analyses become complex.

Example Data

For an example of a cluster randomized trial, a small dataset that is listed in the Appendix is analyzed. These data are a sub–small sample from a study titled Social Capital and Children's Development: A Randomized Controlled Trial (Gamoran, Turley, Turner, & Fish, 2012). The full study was conducted in 52 schools in Phoenix and San Antonio, from 2008 to 2013. The example data include 8 randomly selected Phoenix schools, each with 5 randomly selected students that had third-grade math scores. Although only a tiny fraction of the actual data, the "public use" school and student identifiers are included in the table so the analyses can be replicated. The full public use data file is available at ICPSR (study number 35481).

This study randomly assigned schools (so it is a cluster randomized trial) either to receive a treatment designed to increase the social capital among parents and teachers of first-grade students or to a business-as-usual condition. The intervention is the Families and Schools Together

---

[1] This means using the F distribution, so you may want to review Chapter 2.

(FAST) program. The outcome data for the examples is the third-grade math score of the observed students.[2]

Understanding the Single Level Test as an ANOVA

Multilevel (e.g., mixed, hierarchical) models can trace their origins to multiway ANOVA. Therefore, to prime the discussion of multilevel models, let us briefly review the test of the difference in group means in Chapter 4 as an ANOVA.

One-Way ANOVA F Test for Simple Random Samples

Another method for testing the difference between group means is ANOVA. Many researchers prefer ANOVA to model treatment effects, because it helps extend analyses beyond two groups into more interesting experimental scenarios.

Recall Equation 4.1, where the treatment effect of each group is noted as $\tau_j = \mu_j - \mu$, and the difference between the observations' values of the group means as $e_{ij} = y_{ij} - \mu_j$. This allows us to write the data-generating process as

$$y_{ij} = \mu + \left(\mu_j - \mu\right) + \left(y_{ij} - \mu_j\right), \qquad \text{(Equation 6.1)}$$

which allows us to partition the total squared deviations of each observation from the estimated overall mean (sum of squares total or SST) into the total squared deviations of the estimate of each group's mean from the estimate of the overall mean (weighted by the group size, sum of squares between or SSB) and the total squared deviations of each observation from the estimated group mean (sum of squares within or SSW),

$$\underbrace{\sum_j \sum_i \left(y_{ij} - \bar{y}\right)^2}_{\text{SST}} = \underbrace{n \sum_j \left(\bar{y}_j - \bar{y}\right)^2}_{\text{SSB}} + \underbrace{\sum_j \sum_i \left(y_{ij} - \bar{y}_j\right)^2}_{\text{SSW}}. \qquad \text{(Equation 6.2)}$$

---

[2] Because these are public use data, the raw scores have been collapsed into ordinal (but evenly spaced) categories. For example, 1 represents a score of 11 to 15, 2 represents a score of 16 to 20, and so on, up to a score of 11 that represents a score of 61 or more. The covariate employed is first-grade math ability, which is coded as 1 to 5, with 1 equal to the 10th percentile and 5 being the 90th percentile.

The F ratio employed by one-way ANOVA is a ratio of the estimated variance of the treatment effects to the estimate of the population variance, or the within-group variance. Formally, the numerator of the F ratio, the variance of the treatment effects, is defined as the mean squares between groups (MSB) and is

$$MSB = \frac{n \sum_j \tau_j^2}{p-1} \qquad \text{(Equation 6.3)}$$

with the constraint that $\sum_j \tau_j = 0$. Also recall that τ_j is the difference between the overall mean and the group mean, so we could express Equation 6.3 as

$$MSB = \frac{n \sum_j (\bar{y}_j - \bar{y})^2}{p-1}.$$

The denominator of the F ratio, the variance within groups, is defined as the mean squares within groups (MSW) and is

$$MSW = \frac{\sum_j \sum_i (y_{ij} - \bar{y}_j)^2}{pn - p}. \qquad \text{(Equation 6.4)}$$

This expression should be familiar as the estimate of the population variance (Equation 4.5). The F ratio is then defined as

$$F = \frac{MSB}{MSW}, \qquad \text{(Equation 6.5)}$$

and this test statistic has $p-1$ degrees of freedom in the numerator and $pn - p$ degrees of freedom in the denominator.

These estimates are usually organized in an ANOVA table such as Table 6.1 (see, e.g., Casella, 2008), which clearly lays out the different components of the ANOVA procedure and the partitioning of the total sum of squares (SST) into the between-group sum of squares (SSB) and the within-group sum of squares (SSW). We can compute the F test with the quantities from Table 6.2. In the example data, the difference between groups is 2.35 (see Table 6.7) and n is 20, so the MSB is $\frac{n}{2}(\bar{y}_1 - \bar{y}_0)^2 = \frac{n}{2} 2.35^2 = 55.225$. This is because[3]

$$MSB = \frac{n}{2}(\bar{y}_1 - \bar{y}_0)^2. \qquad \text{(Equation 6.6)}$$

---

[3] This is not obvious, but it can be worked out as follows: If τ_j is estimated with $(\bar{y}_j - \bar{y})$, then $n \sum_j \tau_j^2$ is estimated with (in the two-group case)

$$n\left[(\bar{y}_1 - \bar{y})^2 + (\bar{y}_0 - \bar{y})^2\right].$$

(Continued)

Table 6.1 One-Way ANOVA Table

Source	df	Sum of Squares	Mean Squares	F Test
Between groups	$p-1$	$SSB = n\sum_j \left(\bar{y}_j - \bar{y}\right)^2 = n\sum_j \tau_j^2$	$MSB = \frac{SSB}{p-1}$	$F = \frac{MSB}{MSW}$
Within groups	$pn-p$	$SSW = \sum_j \sum_i \left(y_{ij} - \bar{y}_j\right)^2$	$MSW = \frac{SSW}{pn-p}$	
Total	$pn-1$	$SST = \sum_j \sum_i \left(y_{ij} - \bar{y}\right)^2$		

Note. ANOVA = analysis of variance; SSB = sum of squares between; SSW = sum of squares within; SST = sum of squares total; MSB = mean squares between groups; MSW = mean squares within groups.

Table 6.2 One-Way ANOVA of Third-Grade Math Scores After Treatment

	df	Sum Sq	Mean Sq	F	$\Pr(>F)$
FAST	1.000	55.225	55.225	11.740	0.001
Residuals	38.000	178.750	4.704		

Note. ANOVA = analysis of variance; FAST = Families and Schools Together.

The MSW is 4.704, which is the square of the root MSE in Table 6.7, 2.169. Thus, the F ratio is 11.740. Given (1, 38) degrees of freedom, this leads to $p < 0.001$, which is the same p value as the t test.

(Continued)

Next, note that the estimate of the overall mean can be

$$\bar{y} = \frac{\bar{y}_1 + \bar{y}_0}{2},$$

which makes both $\hat{\tau}_1^2 = (\bar{y}_1 - \bar{y})^2$ and $\hat{\tau}_0^2 = (\bar{y}_0 - \bar{y})^2$ expand to

$$\frac{1}{4}\left(\bar{y}_1^2 + \bar{y}_0^2 - 2\bar{y}_1\bar{y}_0\right).$$

Since there are two τs,

$$MSB = n\left[\frac{1}{2}\left(\bar{y}_1^2 + \bar{y}_0^2 - 2\bar{y}_1\bar{y}_0\right)\right],$$

and since $\bar{y}_1^2 + \bar{y}_0^2 - 2\bar{y}_1\bar{y}_0$ factors into $(\bar{y}_1 - \bar{y}_0)^2$, we arrive at $MSB = \frac{n}{2}(\bar{y}_1 - \bar{y}_0)^2$.

From here, it is simple to show how the F ratio is the square of the t ratio. We can express the t test as

$$t = \frac{\bar{y}_1 - \bar{y}_0}{\hat{\sigma}\sqrt{\frac{2}{n}}}.$$

If we square this and rearrange a little, we have

$$t^2 = \frac{\frac{n}{2}(\bar{y}_1 - \bar{y}_0)^2}{\hat{\sigma}^2},$$

which is the F ratio for the one-way ANOVA. This exercise is also informative in that we confirm that when we have two treatment groups we can retool F statistics into t statistics by taking the square root, as we saw in Chapter 2.

Before we move to cluster trials, we take a moment to explore expected mean squares. In the one-way fixed ANOVA model, we had only one component of variance, the population variance, σ^2. In the multilevel models, we will deal with multiple variance components.

Expected Mean Squares

The idea behind our analyses is that we are testing our data against a null hypothesis that the variance between groups (i.e., the treatment effect) is zero. The F test is a ratio of variances that is a portion of the ratio of expected mean squares. While the derivation of the mean squares will not be presented, readers are encouraged to seek out experimental design texts (an excellent source is Kirk, 1995). For example, the expected mean squares for the one-way fixed ANOVA model are presented in Table 6.3.

The expected mean square between $p = 2$ groups (noted as $E(\text{MSB})$) for a one-way ANOVA (assuming the groups are fixed and not a random

Table 6.3   Expected Mean Squares for One-Way ANOVA Table

Source	Expected Mean Squares	Expected F Test
Between groups	$E(\text{MSB}) = \sigma_e^2 + n\frac{\Sigma_j \tau_j^2}{p-1}$	$F = \dfrac{\sigma_e^2 + n\frac{\Sigma_j \tau_j^2}{p-1}}{\sigma_e^2}$
Within groups	$E(\text{MSW}) = \sigma_e^2$	

Note. ANOVA = analysis of variance; MSB = mean squares between groups; MSW = mean squares within groups.

selection from a larger pool of groups) is essentially the within-group variance plus the variance introduced by the treatment effect,

$$E(\text{MSB}) = \sigma_e^2 + n\frac{\sum_j \tau_j^2}{p-1} = \sigma_e^2 + \frac{n}{2}(\bar{y}_1 - \bar{y}_0)^2. \qquad \text{(Equation 6.7)}$$

The expected mean square within groups is simply the within-group, or population, variance

$$E(\text{MSW}) = \sigma_e^2. \qquad \text{(Equation 6.8)}$$

Thus, the test in expectation is 1 plus the actual test

$$\frac{E(\text{MSB})}{E(\text{MSW})} = \frac{\sigma_e^2 + \frac{n}{2}(\bar{y}_1 - \bar{y}_0)^2}{\sigma_e^2} = 1 + \frac{\frac{n}{2}(\bar{y}_1 - \bar{y}_0)^2}{\sigma_e^2}. \qquad \text{(Equation 6.9)}$$

If the variance introduced by treatment is 0, this expression's expected value is the null hypothesis's ratio of 1. Thus, the test is the ratio in addition to 1, $\frac{\frac{n}{2}(\bar{y}_1 - \bar{y}_0)^2}{\sigma_e^2}$. In the more complex models, we will use expected mean squares to better understand the tests.

As we saw in Chapter 2, we can simply take the square root of this expression to find the t test with $2n - 2$ degrees of freedom. We can do this since an F test with one degree of freedom in the numerator (because $p - 1 = 2 - 1 = 1$, see Table 6.1) is the square of the t test with the denominator's degrees of freedom ($pn - p = 2n - 2$). When making the conversion from an F test with $(1, df_d)$ degrees of freedom to a t test, we always use the denominator's degrees of freedom, df_d, for the t test.

The Hierarchical Mixed Model for Cluster Randomized Trials

In this section, we consider a cluster randomized trial. We begin with specifying the data generating process and then specify the multilevel regression model that is typically used for such analyses. We then find the parameters for power using a two-way mixed ANOVA model.

The reason why we consider these "hierarchical" linear models is because we have perfect nesting of clusters within fixed treatment groups. For example, if we had two treatment groups and two clusters per treatment group, we could think of the structure of our data using random groups nested in the treatments. To make the point concrete, we could say, for example, that the treatment groups are "control" and "treatment" and four schools are nested within them (see Figure 6.1).

Figure 6.1 Visualization of hierarchical nesting of clusters within treatment groups, where treatments are "Control" and "Treatment" and clusters are schools.

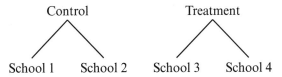

Suppose we have $j = \{1, 2, \ldots p\}$ treatments (in this volume, $p = 2$), each with $k = \{1, 2, \ldots m\}$ clusters assigned to them, and each cluster has $i = \{1, 2, \ldots n\}$ units. The data-generating process for this model adds a term and subscripts to Equation 4.1:

$$y_{ik(j)} = \mu + \tau_j + b_{k(j)} + e_{ik(j)}, \qquad \text{(Equation 6.10)}$$

where $y_{ik(j)}$ is the outcome for the ith unit in cluster k assigned to treatment group j, $\tau_j = \mu_j - \mu$ is the treatment effect for group j (as before), $b_{k(j)} = \mu_{k(j)} - \mu_j$ is a random effect for cluster k net of the treatment effect, and $e_{ik(j)} = y_{ik(j)} - \mu_{k(j)}$ is the within-group and cluster error term. The reason we term $b_{k(j)}$ as a random effect is because we assume that the clusters were picked by some random process (and so we use a Roman letter and not a Greek letter). However, we still assume that we are analyzing all possible treatments, thus τ_j is a fixed effect (noted with a Greek letter). The reason we call these "mixed" models should be clear. Apart from the error term, this model is now a "mix" of fixed and random effects.

Since our data were observed with random processes (randomly obtaining clusters and randomly obtaining units within clusters), we have two sources of variance net of the treatment effect. These variance components are based on the variances of the random effects. Specifically, $e_{ik(j)} \sim N(0, \sigma_e)$ and $b_{k(j)} \sim N(0, \sigma_b)$. Thus, σ_e^2 and σ_b^2 are variance components, and the total variance in the outcome, net of treatment, is $\sigma^2 = \sigma_b^2 + \sigma_e^2$.

Treatment Effect Test Using a Linear Mixed Model or Hierarchical Linear Model

Analysis of cluster randomized trials can be performed with a two-way ANOVA model, but today such analyses are usually performed with a linear mixed model (LMM; McCulloch & Searle, 2001) or hierarchical linear model (HLM; Raudenbush & Bryk, 2002) using an iterative technique such as maximum likelihood or restricted maximum likelihood (details of estimation are beyond the scope of this volume).

LMMs and HLMs are equivalent models that use different notation systems. In the LMM notation system, we add subscripts and random effects to a linear regression such as Equation 4.7 to write the model

$$y_{ik(j)} = \gamma_0 + \gamma_1 T_{k(j)} + b_{k(j)} + e_{ik(j)}. \qquad \text{(Equation 6.11)}$$

In the HLM notation system, we conceive of each "level" of data being represented by different related models. For units within clusters (such as students within schools), we conceive of the "level-1" model as[4]

$$y_{ik(j)} = \pi_{0k(j)} + e_{ik(j)}, \qquad \text{(Equation 6.12)}$$

where $\pi_{0k(j)}$ is the average outcome for units within cluster k and treatment group j and $e_{ik(j)}$ is the within-cluster-group residual distributed normally with a standard deviation of σ_e.

We then posit that each cluster group's mean is a function of a set of covariates (namely, the treatment indicator) and a random effect

$$\pi_{0k(j)} = \gamma_{00} + \gamma_{01} T_{k(j)} + b_{0k(j)}. \qquad \text{(Equation 6.13)}$$

Here, γ_{00} is the overall average of the means for the control group clusters ($T = 0$) and γ_{01} is the difference between the average of treatment group cluster means and the average of the control group cluster means. Finally, $b_{0k(j)} = \pi_{0k(j)} - (\gamma_{00} + \gamma_{01} T_{k(j)})$, or the difference between the cluster-group specific mean and the treatment or control group mean. If we replace $\pi_{0k(j)}$ in Equation 6.12 with Equation 6.13 we find that we arrive at the equivalent of Equation 6.11.

The variance of the random effects are the estimated variance components. The random effect $e_{ik(j)}$ is distributed normally with a mean of zero and standard deviation σ_e, and the random effect $b_{0k(j)}$ is distributed normally with a mean of zero and standard deviation σ_b.

The use of maximum likelihood makes power analysis difficult. However, restricted maximum likelihood, which is used by most software that computes LMM or HLM models, gives estimates analogous to ANOVA models (Raudenbush & Bryk, 2002). Thus, ANOVA's expected mean squares can be helpful in planning studies.

---

[4] Again, I avoid the use of the letter β to avoid within-volume confusion.

Table 6.4   Expected Mean Squares for Two-Way Mixed ANOVA Table Where the Treatment Source Is Fixed but the Cluster Source Is Random

Source	Expected Mean Squares
Between groups	$E(MS_\tau) = \sigma_e^2 + n\sigma_b^2 + nm\frac{\Sigma_j \tau_j^2}{p-1}$
Between clusters	$E(MS_b) = \sigma_e^2 + n\sigma_b^2$
Within	$E(\text{MSW}) = \sigma_e^2$

Mixed Two-Way ANOVA Test of the Treatment Effect in Cluster Randomized Trials

One-way ANOVA splits the variation in an outcome across a single factor. Two-way ANOVA splits the variation in an outcome across two factors. In this case, one factor is the cluster (which is a random sample of clusters) and the other is the treatment group (which is a fixed universe of possible treatment groups). Without getting into the details about the sums of squares and how to compute them, we return to expected mean squares, which can be defined in terms of sample sizes and variance components.

For the mixed hierarchal test, we test the treatment effect with the ratio of MS_τ to MS_b, so the expected test is

$$\frac{E(MS_\tau)}{E(MS_b)}$$

since the components in MS_b (the denominator) can be used to isolate the treatment effect in MS_τ (the numerator; Schultz, 1955). The expected mean squares are defined in Table 6.4 (Kirk, 1995). The expected mean square for MS_τ is (when $p = 2$)

$$E(MS_\tau) = \sigma_e^2 + n\sigma_b^2 + nm\frac{\Sigma_j \tau_j^2}{p-1} = \sigma_e^2 + n\sigma_b^2 + \frac{n}{2}m(\mu_1 - \mu_0)^2$$

(Equation 6.14)

and for the clusters MS_b it is

$$E(MS_b) = \sigma_e^2 + n\sigma_b^2.$$

(Equation 6.15)

Thus, the test statistic is the ratio

$$F = \frac{\frac{1}{2}nm(\mu_1 - \mu_0)^2}{\sigma_e^2 + n\sigma_b^2}.$$

(Equation 6.16)

The t *Test for the Treatment Effect in Cluster Randomized Trials*

The F test (Equation 6.16) can be rearranged and its square root taken to form the analogous t statistic,

$$t = \frac{\mu_1 - \mu_0}{\sqrt{\frac{2}{nm}\left(\sigma_e^2 + n\sigma_b^2\right)}}, \qquad \text{(Equation 6.17)}$$

where the sampling variance of the treatment group difference is

$$\text{vâr}\{\bar{y}_1 - \bar{y}_0\} = \frac{2}{nm}\left(\hat{\sigma}_e^2 + n\hat{\sigma}_b^2\right). \qquad \text{(Equation 6.18)}$$

This t test has $2m - 2$ degrees of freedom (we use the number of clusters, $2m$, and not the total sample size, $2nm$). We use the cluster sample size because in the F statistic we are dividing by the mean squares between clusters (MS_b), and *not* the within-group-cluster mean square.

Power Parameters for Cluster Randomized Trials

Unfortunately, Equation 6.17 is not an ideal formula to work with for power analysis because there are no scale-free parameters. First, we need an effect size. One possibility is to use something like Equation 3.10 where the difference in means is divided by the total standard deviation (Hedges, 2007),

$$\delta = \frac{\mu_1 - \mu_0}{\sigma} = \frac{\mu_1 - \mu_0}{\sqrt{\sigma_b^2 + \sigma_e^2}}. \qquad \text{(Equation 6.19)}$$

Next, with variance components, we can construct another scale-free parameter. The variance between clusters, σ_b^2, is the intraclass covariance of the units within clusters (McCulloch & Searle, 2001). It measures how correlated units within clusters are to each other. We can standardize this metric by the total variation to produce an actual correlation coefficient. This scale-free parameter is the intraclass correlation, ρ_{intra},

$$\rho_{intra} = \frac{\sigma_b^2}{\sigma_b^2 + \sigma_e^2}, \qquad \text{(Equation 6.20)}$$

and it makes power analysis for cluster randomized trials tractable. The interpretation of the intraclass correlation is that it represents the correlation of units within clusters.

If we assume that the total variance is $\sigma^2 = 1$, then we can simply use the difference in means as the effect size. Also, with the intraclass correlation (Equation 6.20), we can transform the variance components into scale-free parameters, $\sigma_b^2 = \rho_{intra}$ and $\sigma_e^2 = 1 - \rho_{intra}$. With this, we can express the t statistic as

$$t = \delta \sqrt{\frac{m}{2}} \sqrt{\frac{1}{\frac{1-\rho_{intra}}{n} + \rho_{intra}}} = \delta \sqrt{\frac{nm}{2}} \sqrt{\frac{1}{1 + (n-1)\rho_{intra}}},$$

where $1 + (n-1)\rho_{intra}$ is the "design effect" (Kish, 1965).

Typically, the "design effect" of a sampling design is the ratio of the sampling variance of an estimate using the appropriate analysis method to the sampling variance of the estimate from an analysis that assumes a simple random sample. In this case, the design effect can be found by taking the ratio of Equation 6.18 to Equation 4.18, if you assume that mn in the clustered design is n in the simple random sample design and that $\sigma_b^2 + \sigma_e^2 = \sigma^2$.

The noncentrality parameter for a cluster randomized trial is

$$\lambda = \underbrace{\delta}_{\text{Effect size}} \underbrace{\sqrt{\frac{nm}{2}}}_{\text{Sample size}} \underbrace{\sqrt{\frac{1}{1 + (n-1)\rho_{intra}}}}_{\text{Design effect}} \qquad \text{(Equation 6.21)}$$

where δ is the effect size, n is the number of units per cluster, m is the number of clusters per treatment, and ρ_{intra} is the intraclass correlation.

The Use of Uncorrelated Covariates in Cluster Randomized Trials

As we saw in the case of single-level models in Chapter 4, using controls that are uncorrelated with the treatment indicator can improve power by reducing conditional variance. This is also true in cluster randomized trials. Suppose we employ a covariate, x, that is uncorrelated with treatment assignment due to randomization. To maximize the benefit of this covariate in a cluster randomized trial, we include both the cluster-mean-centered unit-level value of x and its cluster mean, $\bar{x}_{k(j)}$ in the regression model

$$y_{ik(j)} = \gamma_0 + \gamma_1 T_{k(j)} + \gamma_2 \left(x_{ik(j)} - \bar{x}_{k(j)}\right) + \gamma_3 \bar{x}_{k(j)} + b^*_{k(j)} + e^*_{ik(j)}.$$

$$\text{(Equation 6.22)}$$

I use the asterisk (∗) notation to remind us that the random effects are reduced due to the inclusion of covariates. The variance components are reduced by a factor related to the correlation (within treatment groups) of the covariate to the random effects. For example, the variance of $b_{k(j)}^*$ would be

$$\sigma_b^{2*} = \sigma_b^2 \left(1 - R_{cluster}^2\right),$$

and the variance of $e_{ik(j)}^*$ would be

$$\sigma_e^{2*} = \sigma_e^2 \left(1 - R_{unit}^2\right),$$

where $R_{cluster}^2$ is the proportion of variance at the cluster level that is explained by the cluster-mean covariate and R_{unit}^2 is the proportion of variance explained by the covariate within clusters.

Thus, the variance of the difference (Equation 6.18) becomes

$$\text{vâr}\{\bar{y}_1 - \bar{y}_0\} = \frac{2}{nm}\left(\hat{\sigma}_e^2\left(1 - R_{unit}^2\right) + n\hat{\sigma}_b^2\left(1 - R_{cluster}^2\right)\right).$$

Again, we can use scale-free parameters to convert this into a tractable form. If we again replace $\sigma_b^2 = \rho_{intra}$ and $\sigma_e^2 = 1 - \rho_{intra}$, the variance of the difference can be arranged into this form:

$$\text{vâr}\{\delta\} = \frac{2}{nm}\left(1 + (n-1)\rho_{intra} - \left(R_{unit}^2 + \left(nR_{cluster}^2 - R_{unit}^2\right)\rho_{intra}\right)\right).$$

(Equation 6.23)

This makes the noncentrality parameter (Hedges & Rhoads, 2010)

$$\lambda = \delta\sqrt{\frac{nm/2}{1 + (n-1)\rho_{intra} - \left(R_{unit}^2 + \left(nR_{cluster}^2 - R_{unit}^2\right)\rho_{intra}\right)}}$$

(Equation 6.24)

for a test with $2m - 2 - q$ degrees of freedom, where q is the number of cluster covariates employed in the model.

Example Analysis of a Cluster Randomized Trial

In this section, we return to the example data for this chapter and perform an analysis to make these parameters more understandable. The sample is made up of $n = 4$ students per school, where $m = 5$ schools are assigned each to treatment and control.

Table 6.5 provides some basic summary statistics of the outcome. From the means, it appears that the treatment group scored 2.35 points

Table 6.5 Summary Statistics of Third-Grade Math Scores After Treatment

	N	Mean	Standard Deviation
No program (control)	20.000	6.100	2.532
FAST program (treatment)	20.000	8.450	1.731
Total sample	40.000	7.275	2.449

Note. FAST = Families and Schools Together.

higher on the math assessment. This is confirmed in Table 6.7 with the report of the FAST program coefficient of 2.35. In the oridinary least squares (OLS) model, this results in a t test of $\frac{2.350}{0.686} = 3.426$, which is indeed the square root of the F test reported in Table 6.2.

The "Mixed 1" model provides the estimated test using the mixed model, which results in a t test of $\frac{2.350}{0.762} = 3.084$, a smaller, less significant test. The reason is that the variance of the difference increased by a factor of $\frac{0.762^2}{0.686^2} = 1.234$. This is the design effect, which we can reproduce by calculating the intraclass correlation from the variance components. The school-level variance is 0.351, and the residual within-school variance is 4.408, which corresponds to an intraclass correlation of

$$\rho_{intra} = \frac{0.351}{0.351 + 4.408} = 0.074.$$

With the intraclass correlation, we can compute the design effect as $1 + (n - 1)\rho_{intra} = 1 + (4 - 1)0.074 = 1.222$ (the difference with 1.234 is due to rounding).

Next, we consider a covariate, which is summarized in Table 6.6, where we see that the mean of the pretest for treatment and control is the same. Thus, the pretest is not correlated with the treatment group indicator. In the "Mixed 2" model in Table 6.7, we see that the coefficient for the FAST program is unchanged, which is also expected with

Table 6.6 Summary Statistics of First-Grade Math Scores Before Treatment

	N	Mean	Standard Deviation
No program (control)	20.000	3.500	1.235
FAST program (treatment)	20.000	3.500	1.318
Total sample	40.000	3.500	1.261

Note. FAST = Families and Schools Together.

Table 6.7   Models Predicting Third-Grade Math Scores

	OLS	Mixed 1	Mixed 2
Intercept	6.100***	6.100***	5.956**
	(0.485)	(0.539)	(1.873)
FAST program vs. control	2.350**	2.350**	2.350***
	(0.686)	(0.762)	(0.621)
First-grade math ability			0.946***
			(0.279)
School mean first-grade ability			0.041
			(0.520)
R^2	0.236		
N	40	40	40
Root MSE	2.169		
Number clusters		10	10
σ_b^2		0.351	0.148
σ_e^2		4.408	3.263

Note. First-grade math ability group-mean centered. Standard errors in parentheses. FAST = Families and Schools Together; OLS = ordinary least squares; MSE = mean squared error.
$^*p < 0.05.$ $^{**}p < 0.01.$ $^{***}p < 0.001.$

an uncorrelated covariate. However, the standard error is smaller than in the other models. The reason for this is the covariate, which has reduced the variance components at the school and within-school (residual) level. At the school level, the cluster covariate reduced the between-school variance by $R^2_{cluster} = 1 - \frac{0.148}{0.351} = 0.578$ and the unit-level covariate reduced the within-school variance by $R^2_{unit} = 1 - \frac{3.263}{4.408} = 0.260$. Thus, the design effect here is

$$1 + (n-1)\rho_{intra} - \left(R^2_{unit} + \left(nR^2_{cluster} - R^2_{unit} \right) \rho_{intra} \right)$$

which is

$$1 + (4-1)0.074 - (0.260 + (4 \times 0.578 - 0.260)0.074) = 0.810.$$

This is confirmed by taking the ratio of the model estimate variances $\frac{0.621^2}{0.686^2} = 0.819$ (again, the difference is due to rounding).

Power Analyses for Cluster Randomized Trials

Finding Power A Priori

Once the noncentrality parameter is calculated (Equation 6.24), the power analysis is the same as in the rest of the t tests in this volume. Once λ is calculated, the power of the test can be computed with a computer using the H function introduced in Chapter 4, that is, Equations 4.23 and 4.24. Note that the degrees of freedom for this test are $2m - 2 - q$, where q is the number of cluster level covariates used in the test. If no covariates are used, simply set q, $R^2_{cluster}$, and R^2_{unit} to zero.

Finding the Sample Size

In two-level models there are two sample sizes: the number of clusters and the number of units within clusters. As you will see below, adding units benefits power only to a point, whereas adding clusters provides greater benefit. This section is focused on finding the number of clusters given a number of units within clusters.

Add Units or Clusters?

One question that often arises in planning cluster randomized trials is the trade-off between adding units within clusters or adding clusters. While this is not a straightforward question, a *broad* piece of advice is to add clusters at the expense of units within clusters.[5] To see why this is the case, consider Figure 6.2. In this figure we hold fixed the effect size and intraclass correlation, $\delta = 0.3$ and $\rho_{intra} = 0.2$, and plot the power contours for differing combinations of m and n.

In Figure 6.2, several lines are depicted for areas of power bordering on 0.4 to 0.9 levels of power. The area to the left of the 0.4 curve are combinations of m and n that have power less than 0.4. The area between 0.4 and 0.5 are combinations of m and n that have power between 0.4 and 0.5. For example, if a design has 20 clusters per treatment group and 15 units per cluster (for a total sample size of $2 \times 20 \times 15 = 600$), power is less than 0.5 but greater than 0.4.

What becomes apparent after close study of Figure 6.2 is that increasing the units per cluster impacts power only until about 30 units. That is,

---

[5] However, you will find that different combinations of units and clusters can achieve the same power, and so a major consideration is cost (Raudenbush, 1997).

Figure 6.2 Contour power plot for cluster randomized designs with $\delta = 0.3$ and $\rho_{intra} = 0.2$ by values of m and n.

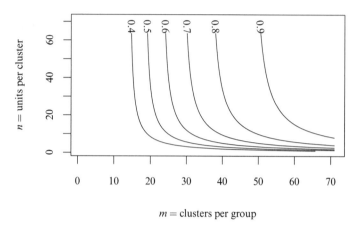

n = units per cluster

m = clusters per group

the power curves do not change much when you look vertically after 30 units. However, the power curves do change when you look horizontally for most of the axis, indicating that it is more powerful to add clusters per treatment than it is to add units per cluster.

Finding the Number of Clusters With Fixed Unit Size

As with the other t tests in this volume, we first start with a standard normal approximation. Rearranging Equation 6.24 we can find the formula for the number of clusters given a number of units per cluster,

$$m_z \approx \frac{2\left(z_{critical} - z_\beta\right)^2 D}{n\delta^2},$$
(Equation 6.25)

where D is the design effect:

$$D = 1 + (n-1)\rho_{intra} - \left(R^2_{unit} + \left(nR^2_{cluster} - R^2_{unit}\right)\rho_{intra}\right).$$
(Equation 6.26)

Once that approximation is in hand, we again refine our estimate using the quantiles of the t distribution

$$m \approx \frac{2\left(t_{(2m_z-2-q)critical} - t_{(2m_z-2-q)\beta}\right)^2 D}{n\delta^2},$$
(Equation 6.27)

where D is the design effect as defined above (Equation 6.26) and q is the number of covariates (used in obtaining the t distribution quantiles).

Example

Suppose we were planning a cluster randomized trial for a new student curriculum. Our test will be a two-tailed test with $\alpha = 0.05$ and we wish to achieve power of 0.8. We plan to randomize schools to treatment conditions, and for each school we expect to have $n = 30$ students. We expect that the treatment will improve math scores by $\delta = 0.5$ population standard deviations. We expect that the intraclass correlation of math scores within schools is about $\rho_{intra} = 0.2$, and that the covariate at the student (unit) level explains about a proportion of $R^2_{unit} = 0.25$ of the variance and about $R^2_{cluster} = 0.64$ at the school (cluster) level. The first step is to calculate the design effect, D:

$$D = 1 + (n-1)\rho_{intra} - \left(R^2_{unit} + \left(nR^2_{cluster} - R^2_{unit}\right)\rho_{intra}\right),$$

$$D = 1 + (30-1)0.2 - (0.25 + (30 \times 0.64 - 0.25)0.2) = 2.76.$$

Next, we use the normal approximation to find an approximate number of clusters needed. That means that $z_{critical} = z_{0.975} = 1.96$ and $z_{.2} = -0.842$. Using these values and the design effect we calculated, we find

$$m_z \approx \frac{2\left(z_{critical} - z_\beta\right)^2 D}{n\delta^2} \approx \frac{2(1.96 - (-0.842))^2 2.76}{30 \times 0.5^2} \approx 5.778$$

or about 6 schools (clusters) per treatment. We then use this number and the t distribution quantiles in the Appendix to refine our estimate, using $2m_z - 2 - q = 2 \times 6 - 2 - 1 = 9$ degrees of freedom ($q = 1$ for the one cluster level covariate). Thus, $t_{(2m_z-2-q)critical} = t_{(9)0.975} = 2.262$ and $t_{(2m_z-2-q)\beta} = t_{(9)0.2} = -0.883$ and

$$m \approx \frac{2\left(t_{(2m_z-2-q)critical} - t_{(2m_z-2-q)\beta}\right)^2 D}{n\delta^2},$$

$$m \approx \frac{2(2.262 - (-0.883))^2 2.76}{30 \times 0.5^2} \approx 7.280,$$

or about 8 schools (clusters) per treatment group for a total of 16 schools. The actual power of this design is higher than 0.8 (0.861) because we rounded the sample sizes up.

Finding the Minimum Detectable Effect

As before, we can rearrange Equation 6.24 to find the formula for the MDES,

$$\delta_m = \left(t_{(2m-2-q)critical} - t_{(2m-2-q)\beta}\right)\sqrt{\frac{2D}{nm}}, \qquad \text{(Equation 6.28)}$$

where D is the design effect as defined above (Equation 6.26) and q is the number of covariates (used in obtaining the t distribution quantiles).

Example

Suppose our resources allowed for only $m = 7$ schools per treatment. We know that since we rounded up in our steps to find the sample size, we may have an acceptable minimum detectable effect with 14 total schools. We can then use the same design effect as before ($D = 2.76$), and then adjust our t distribution quantiles to $t_{(2m_z-2-q)critical} = t_{(11)0.975} = 2.201$ and $t_{(2m_z-2-q)\beta} = t_{(11)0.2} = -0.876$. Again we plan on $n = 30$ students per school. The MDES is

$$\delta_m = \left(t_{(2m-2-q)critical} - t_{(2m-2-q)\beta}\right) \sqrt{\frac{2D}{nm}},$$

$$\delta_m = (2.201 - (-0.876)) \sqrt{\frac{2 \times 2.76}{30 \times 7}} = 0.499,$$

which is an effect size that is close to the 0.5 we desired in the present example.

Summary

In this chapter, we explored multilevel designs, specifically two-level designs in which the cluster (level-2) was the unit of randomization. While the analysis of these studies is usually performed using a mixed regression procedure, power computations for multilevel designs are more approachable when the ANOVA framework is employed. Thus, this chapter includes a brief review of ANOVA and expected mean squares. Using expected mean squares, this chapter produced the noncentrality parameters for the cluster randomized trial. A key parameter in designing cluster randomized trials is the extent to which units within clusters correlate, which is operationalized as the intraclass correlation. The higher the intraclass correlation, the greater the design effect, which is a factor that increases sampling variance. Another result to consider is that it is generally more helpful to add clusters to the sample rather than units within clusters. As always, covariates uncorrelated with the treatment indicator are helpful in reducing sample size requirements.

CHAPTER 7. MULTILEVEL MODELS II: TESTING THE DIFFERENCE IN GROUP MEANS IN TWO-LEVEL MULTISITE RANDOMIZED TRIALS

In contrast to randomizing clusters (Chapter 6), another choice with two-level designs is to randomize units within clusters. This essentially means that each cluster, or "site," becomes its own small trial. This allows researchers to estimate how treatment effects vary across sites. So-called multisite randomized trials are also popular in social and health experiments (Raudenbush & Liu, 2000). In this chapter, we explore power analyses for these designs.

Suppose we have $j = \{1, 2, \ldots p\}$ treatments (in this volume, $p = 2$), each with $i = \{1, 2, \ldots n\}$ units assigned to them within each cluster, and each set of units is assigned to $k = \{1, 2, \ldots m\}$ clusters (sites). The data-generating process for this model is different from Equation 6.10 in that it adds an interaction term and has somewhat different subscripts:

$$y_{ijk} = \mu + \tau_j + b_k + (\tau b)_{jk} + e_{ijk}. \qquad \text{(Equation 7.1)}$$

Here, y_{ijk} is the outcome for the ith unit in treatment group j in cluster k, $\tau_j = \mu_j - \mu$ is the treatment effect for group j (as before), $b_k = \mu_k - \mu$ is a random effect for cluster k, and $(\tau b)_{jk} = \mu_{jk} - \mu_j - \mu_k + \mu$ is the random effect on the treatment effect by cluster, and $e_{ijk} = y_{ijk} - \mu_{jk}$ is the within-group and cluster error term. The reason we term b_k and the interaction $(\tau b)_{jk}$ as random effects is because we assume that the clusters were picked by some random process. However, we still assume that we are analyzing all possible treatments, and thus τ_j by itself is a fixed effect.

This design is visualized in Figure 7.1, where each cluster has a treatment effect. For example, in Figure 7.1 we have four schools, each with a treatment and control group. From this model we have three variance components. First, we note that $e_{ijk} \sim N(0, \sigma_b)$ and that $b_k \sim N(0, \sigma_b)$. Second, we also note that $(\tau b)_{jk} \sim N(0, \sigma_{\tau b})$.

Treatment Effect Test Using a Linear Mixed Model or Hierarchical Linear Model

As with cluster randomized trials, we can estimate a mixed model to perform the analysis to test the treatment effect. The regression model in mixed notation is as follows. The outcome for the ith unit assigned to treatment group j in cluster k, y_{ijk}, is modeled as a function of a treatment

Figure 7.1 Visualization of multisite nesting of treatment groups within clusters, where treatments are "Control" and "Treatment" and clusters are schools.

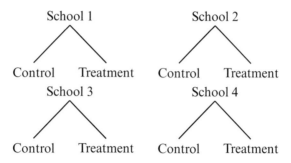

indicator (T), a random effect for the cluster b_k, a random effect of the treatment effect $(Tb)_{jk}$, and a within-group residual, e_{ijk}:

$$y_{ijk} = \gamma_0 + \gamma_1 T_{ijk} + b_k + (Tb)_{jk} + e_{ijk}. \qquad \text{(Equation 7.2)}$$

Exactly how the interaction works and what it means can be better understood with HLM notation. Using HLM notation, we can posit a "level-1" model where the outcome is a function of a cluster-specific control group mean, π_{0k}, a cluster-specific treatment effect, π_{1k}, and a within-cluster residual e_{ijk}:

$$y_{ijk} = \pi_{0k} + \pi_{1k} T_{ijk} + e_{ijk}. \qquad \text{(Equation 7.3)}$$

Next, we have two "level-2" models. The cluster-specific control group mean is a function of the average of cluster control group means plus a cluster-specific random effect,

$$\pi_{0k} = \gamma_{00} + b_{0k}, \qquad \text{(Equation 7.4)}$$

and similarly, the cluster-specific treatment effect is a function of an average of the cluster-specific treatment effects

$$\pi_{1k} = \gamma_{10} + b_{1k}. \qquad \text{(Equation 7.5)}$$

Equation 7.5 is important because it shows how the random effect b_{1k} is the difference between the cluster-specific treatment effect and the average cluster-specific treatment effect. When we replace π_{0k} in

Table 7.1 Expected Mean Squares for Two-Way Crossed ANOVA Table Where the Treatment Source Is Fixed but the Cluster Source Is Random

Source	Expected Mean Squares
Between groups	$E(MS_\tau) = \sigma_e^2 + n\frac{p-1}{p}\sigma_{\tau b}^2 + nm\frac{\Sigma_j \tau_j^2}{p-1}$
Between clusters	$E(MS_b) = \sigma_e^2 + np\sigma_b^2$
Interaction	$E(MS_{\tau b}) = \sigma_e^2 + n\frac{p-1}{p}\sigma_{\tau b}^2$
Within	$E(\text{MSW}) = \sigma_e^2$

Note. ANOVA = analysis of variance; MSW = mean squares within groups.

Equation 7.3 with Equation 7.4 and replace π_{1k} in Equation 7.3 with Equation 7.5, we arrive at the mixed model

$$y_{ijk} = \gamma_{00} + b_{0k} + (\gamma_{10} + b_{1k})\,T_{ijk} + e_{ijk},$$

$$y_{ijk} = \gamma_{00} + b_{0k} + \gamma_{10}T_{ijk} + b_{1k}T_{ijk} + e_{ijk},$$

where $\gamma_{00} = \gamma_0$ in the mixed notation, $\gamma_{10} = \gamma_1$ in the mixed notation, and $b_{1k}T_{ijk} = (Tb)_{jk}$ in the mixed notation.

Again, the variances of the random effects are the estimated variance components. The random effect $e_{ik(j)}$ is distributed normally with a mean of zero and standard deviation σ_e, the random effect $b_{0k(j)}$ is distributed normally with a mean of zero and standard deviation σ_b, and the random effect $(Tb)_{jk}$ is distributed normally with a mean of zero and standard deviation $\sigma_{\tau b}$.

As with cluster randomized trials, these models can be estimated with restricted maximum likelihood. Again, like cluster randomized trials, power analysis is easier when we consider the equivalent ANOVA model.

Mixed Two-Way ANOVA Test of the Treatment Effect in Multisite Randomized Trials

The expected mean squares for this model are presented in Table 7.1. The expected F test of the main treatment effect is the ratio of the main effect to the interaction expected mean squares (Kirk, 1995):

$$\frac{E(MS_\tau)}{E(MS_{\tau b})} = \frac{\sigma_e^2 + n\frac{p-1}{p}\sigma_{\tau b}^2 + nm\frac{\Sigma_j \tau_j^2}{p-1}}{\sigma_e^2 + n\frac{p-1}{p}\sigma_{\tau b}^2}.$$

Note that we use as the denominator the interaction and not the between-clusters mean square. This is because the expected denominator needs to have terms that are in the expected numerator for a valid F test (Brown & Melamed, 1990). Since we know that if $p = 2$ then $n\frac{\Sigma_j \tau_j^2}{p-1} = \frac{n}{2}(\mu_1 - \mu_0)^2$ (Equation 6.6) and $\frac{p-1}{p} = \frac{1}{2}$, the actual F test is

$$F = \frac{\frac{n}{2}m(\mu_1 - \mu_0)^2}{\sigma_e^2 + \frac{n}{2}\sigma_{\tau b}^2}, \qquad \text{(Equation 7.6)}$$

and so the t test is

$$t = (\mu_1 - \mu_0)\sqrt{\frac{nm}{2(\sigma_e^2 + \frac{n}{2}\sigma_{\tau b}^2)}}. \qquad \text{(Equation 7.7)}$$

Power Parameters for Multisite Randomized Trials

Again, we need scale-free parameters to make power analysis possible without knowing details about the metric. One option is to divide both the numerator and denominator of Equation 7.6 by σ_b^2. In doing this, we can arrive at two scale-free parameters. The first is the effect size divided by the square root of the intraclass correlation,

$$\delta_b = \frac{(\mu_1 - \mu_0)}{\sigma_b} = \frac{\delta}{\sqrt{\rho_{intra}}}, \qquad \text{(Equation 7.8)}$$

and a ratio of the variance in the treatment effect to the variance to the variance between cluster means,

$$\upsilon = \frac{\sigma_{\tau b}^2}{\sigma_b^2}. \qquad \text{(Equation 7.9)}$$

Combining these elements (Equations 7.8 and 7.9) with the assumptions that $\sigma_e^2 = 1 - \rho_{intra}$ and $\sigma_b^2 = \rho_{intra}$, we can find a scale-free noncentrality parameter (Hedges & Rhoads, 2010)[1]

$$\lambda = \delta_b\sqrt{\frac{nm}{2\left(\frac{1-\rho_{intra}}{\rho_{intra}} + \frac{n}{2}\upsilon\right)}} = \delta\sqrt{\frac{nm}{\rho_{intra}\left(2\left(\frac{1-\rho_{intra}}{\rho_{intra}} + \frac{n}{2}\upsilon\right)\right)}}$$

---

[1] Note that Hedges and Rhoads use a similar ratio, ω, that is half the size of υ, and thus, $\omega = \upsilon/2$.

and so the noncentrality parameter is

$$\lambda = \underbrace{\delta}_{\text{Effect size}} \underbrace{\sqrt{\frac{nm}{2}}}_{\text{Sample size}} \underbrace{\sqrt{\frac{1}{1 + \left(\frac{n}{2}v - 1\right)\rho_{intra}}}}_{\text{Design effect}}. \qquad \text{(Equation 7.10)}$$

The Use of Uncorrelated Covariates in Multisite Randomized Trials

Equation 7.10 is approachable for handling covariates. Suppose we wished to use uncorrelated covariates in our models, such as a pretest and its cluster mean. We could specify a model such as

$$y_{ijk} = \gamma_0 + \gamma_1 T_{ijk} + \gamma_2 \left(x_{ijk} - \bar{x}_k\right) + \gamma_3 \bar{x}_k + b_k^* + (Tb)_{jk}^* + e_{ijk}^*.$$

$$\text{(Equation 7.11)}$$

I again use the asterisk ($*$) notation to remind us that the random effects are reduced due to the inclusion of covariates. The variance components are reduced by a factor related to the correlation (within treatment groups) of the covariate to the random effects. For example, the variance of $e_{ik(j)}^*$ would be

$$\sigma_e^{2*} = \sigma_e^2 \left(1 - R_{unit}^2\right),$$

and the variance of $(Tb)_{jk}^*$ would be

$$\sigma_{\tau b}^{2*} = \sigma_{\tau b}^2 \left(1 - R_{treat}^2\right),$$

where R_{unit}^2 is the proportion of the within-cluster variance explained by the the covariate and R_{treat}^2 is the proportion of the variance in the treatment effect explained by the cluster mean of the covariate. We can incorporate the effect of covariates into the noncentrality parameter as follows (Hedges & Rhoads, 2010):

$$\lambda = \delta \sqrt{\frac{nm/2}{1 + \left(\frac{n}{2}v - 1\right)\rho_{intra} - \left(R_{unit}^2 + \left(\frac{n}{2}v R_{treat}^2 - R_{unit}^2\right)\rho_{intra}\right)}},$$

$$\text{(Equation 7.12)}$$

where the test has $m - 1 - q$ degrees of freedom and q is the number of cluster level covariates.

Table 7.2 Models Using Simulated Multisite Trial Data

	Mixed 1	Mixed 2
Intercept	55.623***	21.435
	(5.886)	(43.852)
Treatment vs. control	11.340	11.340*
	(7.642)	(5.188)
Covariate		0.509***
		(0.152)
Site covariate mean		0.684
		(0.872)
N	40	40
Number of sites	5	5
σ_b^2	137.670	73.153
$\sigma_{\tau b}^2$	220.834	74.786
σ_e^2	142.269	119.600

Note. Covariate group-mean centered. Standard errors in parentheses.
$^*p < 0.05.$ $^{**}p < 0.01.$ $^{***}p < 0.001.$

Example Analysis of a Multisite Randomized Trial

In this section, we briefly analyze some simulated data.[2] These data are available in the Appendix. The data are simulated achievement scores for units (students; on a scale of 0 to 100) that are possibly impacted by a treatment assigned within sites (schools). The data contain $n = 4$ students in each of $p = 2$ treatment groups in each of the $m = 5$ schools. The results of two mixed models are presented in Table 7.2.

In the first model (without covariates) presented in Table 7.2, we see that students in the treatment condition increase their scores by 11.34 points on average, but the effect is not statistically significant. We can replicate the t test using the parameters reported in the table. The effect size is

$$\delta = \frac{11.340}{\sqrt{137.670 + 142.269}} = 0.678,$$

---

[2] Publicly available multisite trial data proved difficult to obtain.

the υ parameter is

$$\upsilon = \frac{220.834}{137.670} = 1.604,$$

and the intraclass correlation is

$$\rho_{intra} = \frac{137.670}{137.670 + 142.269} = 0.492.$$

This makes the t test (and noncentralty parameter)

$$t = 0.678\sqrt{\frac{4 \times 5}{2}}\sqrt{\frac{1}{1 + \left(\frac{4}{2}1.604 - 1\right)0.492}} = 1.484,$$

which is confirmed by taking the ratio of the coefficient to its standard error $\frac{11.340}{7.642} = 1.484$.

In the second mixed model (with uncorrelated covariates) presented in Table 7.2 we note that the effect of treatment is the same, but is now statistically significant by convention, since its standard error has decreased. We also note that the variance of the cluster intercepts and cluster treatment effects have also decreased. The variance of the unit variance intercepts has decreased by a factor of $R^2_{unit} = 1 - \frac{119.6}{142.269} = 0.159$ and the variance in the treatment effect has decreased by a factor of $R^2_{treat} = 1 - \frac{74.786}{220.834} = 0.661$. Thus, the t test and noncentraltity parameter is now

$$t =$$

$$0.678\sqrt{\frac{4\times5/2}{1 + \left(\frac{4}{2}1.604 - 1\right)0.492 - \left(0.159 + \left(\frac{4}{2}1.604 \times 0.661 - 0.159\right)0.492\right)}}$$

$$= 2.186,$$

which is confirmed with the ratio of the coefficient to its standard error $\frac{11.340}{5.188} = 2.186$. The use of uncorrelated covariates improved the precision of the effect estimate and it is now significant.

Power Analyses for Multisite Randomized Trails

Finding Power A Priori

Once the noncentrality parameter is calculated (Equation 7.12), the power analysis is the same as in the rest of the t tests in this volume. Once λ is calculated, the power of the test can be computed with a computer

Figure 7.2 Contour power plot for multisite randomized designs with $\delta = 0.3$, $\rho_{intra} = 0.2$, and $\upsilon = 0.5$ by values of m and n.

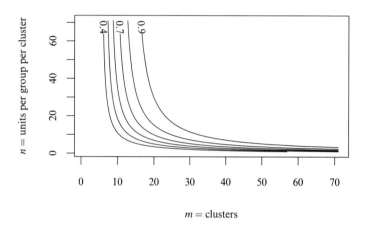

m = clusters

using the H function introduced in Chapter 4, that is, Equations 4.23 and 4.24. Note that the degrees of freedom for this test are $m - 1 - q$, where q is the number of cluster level covariates used in the test. If no covariates are used, simply set q, R^2_{unit}, and R^2_{treat} to zero.

Finding the Necessary Sample Size

Again, as with cluster randomized designs, the improvement in power due to adding units has diminishing effects. However, note that here n is for units per treatment group per cluster. Thus, looking at Figure 7.2, while the contour lines start to become vertical lines after $n = 30$, that means that the total number of units per cluster is 60. That means that adding units to clusters has more purchase in multisite randomized trails compared with cluster randomized trials. Be that as it may, we will still focus on finding the necessary number of clusters.

Finding the necessary number of clusters in multisite randomized trials is similar to the process for cluster randomized trials. We first employ a normal distribution approximation using a formula similar to Equation 6.25,

$$m_z \approx \frac{2 \left(z_{critical} - z_\beta \right)^2 D}{n\delta^2},$$

(Equation 7.13)

where D is a design effect that is different from that in a cluster randomized trial:

$$D = 1 + \left(\frac{n}{2}\upsilon - 1\right)\rho_{intra} - \left(R^2_{unit} + \left(\frac{n}{2}\upsilon R^2_{treat} - R^2_{unit}\right)\rho_{intra}\right).$$
(Equation 7.14)

Once that approximation is in hand, we again refine our estimate using the quantiles of the t distribution:

$$m \approx \frac{2\left(t_{(m_z-1-q)critical} - t_{(m_z-1-q)\beta}\right)^2 D}{n\delta^2}.$$
(Equation 7.15)

Note that we use $m - 1 - q$ degrees of freedom and not $2m - 2 - q$.

Example

For example, suppose we are planning a study with 0.8 power in which we expect an effect size of about 0.25 using a two-tailed test with $\alpha = 0.05$, assuming an υ ratio of about 0.25, an intraclass correlation of about $\rho_{intra} = 0.2$, where a covariate will be employed that explains about a proportion of $R^2_{unit} = 0.5$ of the unit-level variance and about $R^2_{treat} = 0.35$ of the treatment variance. If we planned to have about $n = 50$ units per cluster, the design effect would be

$$D = 1 + \left(\frac{50}{2} \times 0.25 - 1\right) \times 0.2 - \left(0.5 + \left(\frac{50}{2} \times 0.25 \times 0.35 - 0.5\right)0.2\right),$$

which equals 1.213. The normal approximation for the number of clusters is then

$$m_z \approx \frac{2 \times (1.96 - (-0.842))^2 \times 1.213}{50 \times 0.25^2} = 6.095,$$

or about 7 clusters. We then use this in the next approximation using the t distribution with $7 - 1 - 1 = 5$ degrees of freedom, and so $t_{(m_z-1-q)critical} = t_{(7-1-1)0.975} = 2.571$ and $t_{(m_z-1-q)\beta} = t_{(7-1-1)0.2} = -0.92$. We then find that

$$m \approx \frac{2 \times (2.571 - (-0.92))^2 \times 1.213}{50 \times 0.25^2} = 9.461,$$

or about 10 clusters. The power of this test is actually about 0.88, and the power with 9 clusters is about 0.83. This shows that with small samples, the approximation method presented here generally errs on the side of a larger sample.

Finding the Minimum Detectable Effect Size

As before, we can rearrange to find the formula for the MDES,

$$\delta_m = \left(t_{(m-1-q)critical} - t_{(m-1-q)\beta}\right)\sqrt{\frac{2D}{nm}}, \qquad \text{(Equation 7.16)}$$

where D is the design effect as defined above (Equation 7.14) and q is the number of covariates (used in obtaining the t distribution quantiles).

Example

Suppose our resources allowed for only $m = 7$ clusters. We can ask what the minimum detectable effect would be. We can use the same design effect as before ($D = 1.213$), and then adjust our t distribution quantiles to $t_{(m-1-q)critical} = t_{(5)0.975} = 2.571$ and $t_{(m-1-q)\beta} = t_{(5)0.2} = -0.92$. Again, we plan on $n = 50$ students per school. The MDES is

$$\delta_m = (2.571 - (-0.92))\sqrt{\frac{2 \times 1.213}{50 \times 7}} = 0.291,$$

which is an effect size that is close to the 0.25 we desired.

Summary

In this chapter, we continued to explore multilevel designs, specifically two-level designs where the unit (level-1) was the unit of randomization. While the analysis of these studies is usually performed using a mixed regression procedure, power computations for multilevel designs are more approachable when the ANOVA framework is employed. Using expected mean squares, this chapter produced the noncentrality parameters for the multisite randomized trial. Comparing Figure 6.2 with Figure 7.2 we see that multisite randomized trials have more power for the same sample size compared with cluster randomized designs. However, multisite designs risk contamination and other difficulties when there are a mix of treatment and control in the same organization (Bloom, 2005). Thus, cluster randomized trials, which require more clusters, may yield larger effect sizes to compensate for lower power.

CHAPTER 8. REASONABLE ASSUMPTIONS

At this point in the volume, the general approach to power analysis should be clear. Each type of power analysis (finding power, finding sample size, finding effect size) makes assumptions about part of the puzzle to compute the target assumption. To find power, we assume an effect size and sample size. To find the sample size, we assume a level of power and an effect size. To find (the minimum) effect size, we assume power and a sample size. For all these, we also always assume a level of Type I error and type of test (but usually this is the uncontroversial $\alpha = 0.05$ with two tails). In more complex designs, we must assume more parameters. For ANCOVA models, we need to assume something about the correlation of the covariate and the outcome. For clustered samples, we need to assume something about the intraclass correlation. Thus, power analysis is all about assumptions.

We learn as children that assumptions can make people look foolish. This makes power analysis a delicate practice, as it is based entirely on assumptions. To help the reader avoid looking foolish, this chapter presents some strategies for validating assumptions.

While it is not always possible to get a power analysis "right," the goal should be to make it defensible as it is often used in applications for research funding or a contract. As a writer and reviewer, I have often noticed that while no grant has been won *because* of a great power analysis, many have been passed over because of a bad power analysis.

In this chapter, we explore strategies to form reasonable assumptions about design parameters. This is not intended to be a definitive guide to this process, as it depends heavily on the research context. However, the hope is that this set of commonsense practices will help researchers avoid common pitfalls.

Power Analyses Are Arguments

Like much of what researchers do, power analysis is an argument based on a set of premises. In the context of arguing that a piece of research is worth the cost (e.g., in terms of money or risk), we are making the argument that it is likely to succeed. "Success" can mean many things, of course, but let us suppose for a moment that it means achieving an answer to an important question that is reasonably definitive. Power analysis is essential for this, because it provides an estimate of the chance of detecting an effect, should it actually exist.

Power analysis arguments surround two elements of design that form the premises. First, the *sample design* details the type of sample and quantifies all the elements of the sample. For "simple random samples," the important quantity is the total sample size. Sample design details relate to the type of analysis as well. For the test comparing two groups' means, the sample design also involves whether the sample is balanced or not balanced. More complex samples involve more complex quantities. Cluster samples have a sample size for clusters and a sample size within each cluster. If clusters are grouped, then there is the number of clusters within each group, and so on. Most of the sample design is under the control (or constraints) of the research team.[1]

The second set of power analysis arguments surround *design parameters*, such as the expected effect size and correlations. The sampling design, described above, specifies what design parameters are necessary to know in order to evaluate power. For example, in simple random samples, intraclass correlations are not necessary, but effect sizes are necessary if the analysis is to compare two or more groups.

Unfortunately, many researchers fall into the trap of holding to a sampling design (especially sizes) and finding design parameters to support the sample design. This is akin to selecting evidence only if it supports a particular point of view. This often leads to unreasonable expectations that the effect size or correlations are large, when in fact they might be small.

"What Is Large?"

Of course, "small" and "large" are relative terms, and the values associated with them vary widely from discipline to discipline, intervention to intervention, and context to context. It used to be the case that many researchers relied on the so-called "shirt sizes" proposed by Cohen (1992).[2] Such a reliance on prespecified quantities from one broad field of science

---

[1] Other elements of sampling design that are not covered by this volume include sampling weights and stratification by groups. Readers are advised to seek out resources such as Lohr (2009) or the more technical volume by Wolter (2007) for the implications of these aspects of sampling.

[2] These "small," "medium," and "large" effect sizes were based on social data, and even then were tied to the impact of specific interventions.

in the past century is dangerous. Cohen himself expressed that blindly following these conventions was not helpful for larger science. The reason this is dangerous is that using an arbitrary number bypasses important questions about what constitutes a meaningful difference relative to the population variation. While some software still supports pre-filled fields with such shirt sizes, they should be ignored or used only as a last resort with an admission of ignorance.

Increasingly, studies are conducted specifically to estimate naturally occurring effect sizes to benchmark likely changes that can occur from an intervention. One example is Hill et al. (2008), which estimated the effect size associated with a year's worth of schooling. Of course, this depends on several factors, including age and the outcome. In reading, the growth from kindergarten to first grade (a year of schooling) is about 1.5 standard deviations and in math it is 1.14. Later on, from 11th to 12th grade reading changes by about 0.06 standard deviations and math changes by about 0.01 standard deviations (Hill et al., 2008). This study is quite useful for education interventions because it allows for a sense of scale in education research, giving an upper bound to possible effect sizes.[3]

Suppose a researcher proposes a grant to evaluate a curriculum to increase academic achievement. He or she argues that the expected effect size is 0.1. This is "small" by many standards and may seem reasonable. If the researcher is using this intervention for reading on kindergartners, where an increase of 0.1 standard deviations represents about 7% of a year's growth, an argument could be made that this is reasonable. If, instead, the proposed intervention is targeted at 11th-grade math students, the "small" effect size of 0.1 is now 10 times the amount of a year's growth, which may not be reasonable to assume.

Reasonable assumptions are the bedrock of a reasonable power analysis. The best option for gaining reasonable expectations is to use data from a previous study or pilot data, as this will best approximate the conditions for the proposed study. In many situations, having such data will not be the case, and so you must turn to the literature. The rest of this chapter offers some guidance on how to use previous studies for power analysis assumptions.

---

[3] With, of course, the assumption that it is unlikely for an intervention to be worth a year of schooling or more.

Strategies for Using the Literature to Make Reasonable Assumptions

In this section are some strategies for using the literature to make assumptions. Much of this is more art than science, and it involves a lot of judgment on the part of the researcher. A key skill in such searches is the ability to gather important statistics from the studies to calculate the design parameters of interest. Many of the skills in this section have been developed by meta-analysts (Borenstein, Hedges, Higgins, & Rothstein, 2009). In fact, meta-analyses are a good place to start for gathering effect sizes or correlations.

Standardized Differences in Means and Correlations

Estimating the correlation between two variables, whether it is a difference in group means or a linear correlation, is the most important assumption in any power analysis for a study that seeks to uncover relationships between variables. Even when estimating the minimum detectable effect, having a notion as to what is probable is important for comparison. It is always worthwhile to look at the literature to find evidence that relates to the outcome or intervention, or both. This avoids the pitfalls of using "shirt sizes" by examining the size of the effects that are likely within a field and for a given population.

Unfortunately, unless the aim of the research is replication, researchers are unlikely to find a study that has used a similar intervention on a similar outcome.[4] Thus, when looking at other experiments, there are two questions to balance when thinking about the results in the literature:

1. How similar is the independent variable?

2. How similar is the dependent variable?

The guidance on design parameters will probably be from a search that combines studies that either have a similar intervention (which can give a sense as to how effective an intervention can be) or have a similar outcome (which can give a sense as to how much this outcome can change due to an intervention). Both angles are useful in determining an effect

---

[4] If the study is a replication, the strategies offered here still apply: Calculate the effect size if the study does not provide one.

size. Unfortunately, which angle should be weighted more is, again, more art than science.

Meta-analyses are a good place to start, as they often report effect sizes. When meta-analyses are not available, a literature search can reveal studies of either similar interventions or outcomes. Once studies have been identified, we can use the formulas for the effect size to extract the design parameters of interest. If an article provides the sample size, mean, and standard deviation of each group, we can calculate a standardized mean difference. If articles provide correlations instead of standardized mean differences, I provide a conversion formula below. Practically, researchers should find as many studies as possible to assemble their assumptions. Here, I use only two studies, but I do this only for brevity. In practice: look far and wide.

Example

As an example, consider the following. There is a growing literature on the effect of police body-worn cameras (BWCs) on many aspects of policing (see, e.g., White, 2014). Suppose a research team plans to use BWCs with police officers to see if they increase citizen feelings of police legitimacy. Assuming no such intervention has been conducted and published,[5] researchers are left with looking at two aspects of the literature: studies that use the intervention (BWCs) or studies that examine the outcome (police legitimacy).

One study used randomly assigned BWCs in Florida to test effects on serious complaints against officers (which can be argued to be related to feelings of legitimacy; see Jennings, Lynch, & Fridell, 2015). In this study, there were $n_0 = 43$ officers in control and $n_1 = 46$ officers in treatment. The mean number of complaints for control was 0.19 with a standard deviation of 0.39.[6] The mean number of complaints for treatment was 0.09 with a standard deviation of 0.28. Recall that effect sizes are a scale-free parameter that is a standardized difference in the group means,

$$\delta = \frac{\mu_1 - \mu_0}{\sigma}.$$

---

[5] As of this writing, I know of no such randomized trial.

[6] It is obvious that this is not a normally distributed variable, and that the analysis should be a generalized model. However, it is often difficult to find the perfect study. Issues like this often arise in literature searches and so this is a realistic exercise.

We can calculate the pooled standard deviation, which is an estimate of σ, as the square root of Equation 4.6,

$$\hat{\sigma} = \sqrt{\frac{(43-1)0.39^2 + (46-1)0.28^2}{43+46-2}} = 0.34.$$

Since the difference was about 0.1 complaints, the standardized mean difference effect size is about $\delta = 0.1/0.34 = 0.29$.

Another study evaluated an intervention in Australia to see if giving officers a prespecified script to increase feelings of procedural justice had an effect on how citizens felt about their police encounters (Mazerolle, Antrobus, Bennett, & Tyler, 2013). The officers were randomly assigned to a script designed to increase feelings of legitimacy or to a "business-as-usual" condition and encountered citizens at prespecified roadblocks. While the article used a more sophisticated analysis, it did provide the important information to determine a Cohen's d effect size. The mean of the legitimacy scale (which ranged from 1 to 5) for treatment surveys was 4.21, and for control surveys it was 3.96.[7] The standard deviation for both groups was about 0.72.[8] Thus, the effect size was about $(4.21 - 3.96)/0.72 = 0.35$.

In this example, we found the effect size of an intervention on feelings of legitimacy to be about 0.35, and an effect size of BWCs on complaints to be about 0.29. There is no clear answer as to which number to use as an effect size. However, having both numbers so close to 0.3, in this instance, leads us to think that a reasonable effect size for our hypothetical experiment may be 0.3.[9] Of course, a more exhaustive literature search or meta-analysis would be helpful in determining a reasonable effect size.

Converting Correlations to Differences in Means

Often, results from observational research or experiments provide correlations to summarize the effect of a treatment. In other cases, an intervention may seek to capitalize on a variable that has been shown to be correlated with an outcome. In these cases, it is often useful to convert

---

[7] Here, the unit of analysis appeared to be surveys, which were nested within officers. The appropriate analysis should take this into account. However, even though the research team did not, the effect of not using a multilevel model on the effect size is likely minimal.

[8] If they were different, we should calculate the pooled standard deviation using the sample size of each group.

[9] This, of course, was an arbitrary pick, but a round number.

a correlation into a Cohen's d effect size. The meta-analysis literature is helpful here, as meta-analysis often combines effects from studies that are presented in different metrics (Borenstein et al., 2009).[10]

To convert a correlation (ρ) to a standardized mean difference (such as Cohen's d or Hedges's g), we can use the following formula (Borenstein et al., 2009):

$$\delta = \frac{2\rho}{\sqrt{1-\rho^2}}. \qquad \text{(Equation 8.1)}$$

For example, if an article cites a correlation between a treatment variable and an outcome to be 0.18, as was done in the Australian example (Mazerolle et al., 2013), we can convert that into a standardized mean difference with

$$\frac{2 \times 0.18}{\sqrt{1-0.18^2}} = 0.4.$$

Thus, a correlation of 0.18 is equivalent to a mean difference of about 0.4 standard deviations. As you can see, this result is different than the 0.35 we computed from the summary statistics. It is different because these formulas simplify complex relationships and should be used to give a "sense" of possible effect sizes. The reverse conversion formula is (Borenstein et al., 2009)

$$\rho = \frac{\delta}{\sqrt{\delta^2 + a}}$$

where a is a function of the sample sizes in each group,

$$a = \frac{(n_0 + n_1)^2}{n_0 n_1}.$$

If the study is balanced, $a = 4$. For example, we can convert the standardized mean difference from the police legitimacy study using the sample sizes provided in the article ($n_0 = 1,649$, $n_1 = 1,097$) and the effect size of 0.35:

$$\frac{0.35}{\sqrt{0.35^2 + \frac{(1649+1097)^2}{1649 \times 1097}}} = 0.17.$$

This is close to the article's reported correlation of 0.18.

---

[10] Other conversions are also available; for example, we can convert odds ratios to correlations.

Complex Sample Design Parameters

Increasingly, studies involve complex samples that break the assumptions of basic analyses. The most common case is the clustered sample,[11] where the effect of interest is from a variable at the cluster level (such as a cluster randomized trial) or within clusters (such as a multisite randomized trial).

In Chapter 6, we learned that the difference in the sampling variances of the effect can be summarized with "design effects." The design effect for a cluster randomized trial, for example, is detailed in Equation 6.21,

$$D = 1 + (n - 1)\rho_{intra},$$

where n is the number of units per cluster and ρ_{intra} is the intraclass correlation (which is a measure of how correlated units are within a cluster). For example, if $\rho_{intra} = 0.1$ and there are 30 units in each cluster, the variance of the difference in group means increases by a factor of $1 + (30 - 1)0.1 = 3.9$. This means that the standard error (the square root of the variance) also increases, leading to an increase in the Type I error (i.e., less significant).

Chapter 7 also detailed the design effect of multisite randomized trials (where the effect is from a variable within clusters). Here, the design effect also involves the variance in the treatment effect across clusters in addition to the intraclass correlation. The design effect is detailed in Equation 7.10,

$$D = 1 + \left(\frac{n}{2}\upsilon - 1\right)\rho_{intra},$$

where υ is the ratio of the variance in the treatment effect across clusters to the variance of cluster means (see Chapter 7).

Planning for studies that use complex samples typically starts with the simple random sample solution and then applies the appropriate design effect. For example, if the expected t test statistic (and thus the noncentrality parameter) is 2.5 using a simple random sample, but the design effect from the planned cluster sample is expected to be 3.9, we must adjust the noncentrality parameter to be $2.5/\sqrt{3.9} = 1.27$. This drastically decreases power because the noncentrality parameter becomes

---

[11] Of course, there are other types of complex samples that include stratified samples and stratified cluster samples, among others. Many complex samples also use probability weights to account for unequal selection probabilities. Each of these elements creates a design effect on analyses (Lohr, 2009). Stratification typically reduces variances while weights and clusters increase variances.

much smaller. Thus, it is imperative to plan for design effects in complex samples.

While design effects are affected by the within-cluster sample size (n) that is under the control of the researcher, they are also influenced by the intraclass correlation (ρ_{intra}) and treatment heterogeneity parameters (e.g., υ). The latter parameters are often unknown.

Scholars in health and education research have worked to catalog estimates of population intraclass correlations for use in planning studies. Examples include studies from community health surveys (Gulliford, Ukoumunne, & Chinn, 1999), school studies of substance use (see, e.g., Murray et al., 1994), education experiments (Schochet, 2008), national surveys of academic achievement (Hedges & Hedberg, 2007), and state education data system records of achievement (Hedberg & Hedges, 2014; Hedges & Hedberg, 2013; Westine, Spybrook, & Taylor, 2013). Many of these articles also include R^2 values associated with common covariate sets so that researchers can plan on the effectiveness of using pretests and demographic controls.

In health, Murray and colleagues (1994) produced a set of intraclass correlations related to how teen smoking is clustered in schools to be used in future cluster randomized trials of smoking. Table 6 in Murray et al. (1994) provides intraclass correlations for cigarettes smoked per week across a variety of studies representing different school grade levels. For Grades 11 to 12, the intraclass correlation was 0.076, meaning that those planning a study where $n = 30$ can expect a design effect of $1 + (30 - 1)0.076 = 3.204$. This design effect indicates that when compared with a simple random sample, the sampling variances of the expected effects will be more than three times larger.

Education is another area of study in which such design parameters are widely available. Over the past 10 years, Hedges (2007, 2013) and Hedberg (2014), Westine, Spybrook, and Taylor (2013), and others have produced a set of articles detailing intraclass correlations as they relate to academic achievement across a variety of grades and subjects. For example, the intraclass correlation for kindergarten math scores nationally is 0.243 (see Table 2 in Hedges & Hedberg, 2007). This means that those planning a study in which $n = 30$ can expect a design effect of $1 + (30 - 1)0.243 = 8.047$. This design effect indicates that when compared with a simple random sample, the sampling variances of the expected effects will be more than eight times larger.

Less is known about the heterogeneity of effects across sites, and thus papers detailing values for parameters such as υ are scarce. Often, to estimate this parameter, it is useful to have some data on hand.

Summary

Admittedly, this chapter poses more problems with power analysis than answers. Namely, power analysis is based almost entirely on assumptions such as effect size and within-cluster correlation. While researchers can never predict the parameters of future data with certainty, the message of this chapter is that it is never a good idea to guess about these assumptions, and using conventions counts as guessing. The better practice is to search the literature for clues as to what to expect. If there is a meta-analysis that exists about your topic, that will be a good resource. In the likely event that no such study exists, evidence must be pieced together from various sources. A definitive answer will not be reached, but the answer will at least be informed.

CHAPTER 9. WRITING ABOUT POWER

In this chapter, writing about power is addressed. The aim is to give a sense as to what is necessary to convey the careful power analysis that was performed. Often, grant applications or contract proposals have limited space for power analyses, and thus care must be taken to address everything needed by reviewers to judge the proposed design in only a couple of paragraphs.

What to Include

Chapter 8 made the point that power analyses are arguments and should be based on reasonable assumptions. Of course, the application section on power analysis should include all the assumptions. However, this is not sufficient, as a power analysis must also show that the power to detect the effect was computed in a way that is appropriate for the research design. For example, it is unwise to detail a power analysis for a simple random sample when the design calls for a clustered sample. Below are some suggestions about what to include in your power analysis write-up.

Elements of a Power Analysis Section

The power analysis write-up will be different depending on the type of power analysis used (i.e., finding the power, finding the sample size, or finding the minimum detectable effect). In general, it should include the following elements:

1. The sample design proposed (e.g., simple random sample, cluster sample, etc.) and the broad structure of the data; be sure to include factors such as attrition or nonresponse into your design for the final sample size
2. A statement about the statistical model that will be used to estimate the effect and whether covariates are employed
3. The design parameter assumptions used in the power analysis, including citations and/or data to support them; be sure to include the Type I error assumed (α) and whether the test is one or two tailed (if applicable)
4. Citations for the formula (include expression and/or page number) and/or software (include the procedure or command) used to perform the power analysis and how they relate to the statistical model

proposed; if a general statistical package is used for the power analysis, an appendix with the code is often useful

5. This is a good place for the results of the analysis:

If the analysis seeks to estimate the power, given a sample and effect size, then a statement about the achieved power should be added

If the analysis seeks to estimate the sample size, given a level of power and effect size, then a statement relating the sample size necessary and how it matches the proposed budget and recruitment prospects should be added

If the analysis seeks to estimate a minimum detectable effect, given a sample size/design and level of power, then a statement about how this effect size compares with the literature should be added

6. A statement about the sensitivity of the results to deviations from the design parameter assumptions or sampling design assumptions; if space permits, graphs are helpful

7. (Optional, but becoming more important as external validity becomes more of a focus) A statement about how the population(s) used for the assumptions is (are) a good fit for the population to be studied

Examples

In this section, I provide some example language that includes all the elements described above. The first examples are for a simple random sample analysis described in Chapter 4. The next example is for a more complex sample described in Chapter 6. These examples should be used only as examples, as different funding agencies, clients, journals, professional associations, or institutional review boards may have specific guidelines about how to structure the power analysis write-up.

For the examples, we return to the BWC study that we explored in Chapter 8. We are planning a study to examine the effects of BWCs on feelings of police legitimacy. Elsewhere in the application we detail the data collection process and the analysis. Both examples are simplified plans, of course. A study like this would be far more complicated, having to take into account squads and patrol areas, for example.

Simple Random Samples

For these examples, suppose that surveys are distributed to citizens who encounter 10 officers either wearing (treatment) or not wearing (control) a BWC. For each officer, the research team has determined that about 25 surveys will be mailed back during the study period. The surveys are mailed anonymously to the research team and entered into a database. A simple *t* test of the resulting police legitimacy score, comparing treatment and control, will be performed. For simplicity, we assume that the sample will be balanced.

Power Given an Effect Size and Sample Size

Here is a possible section outlining the achieved power:

We plan on utilizing 20 officers from the City Police Department and randomizing $n = 10$ officers to wear a BWC (treatment) and $n = 10$ officers to not wear a BWC (control). The officers will hand out about 100 surveys to citizens during the study period, and we expect to receive approximately 25 percent of them for a total sample of $2 \times 10 \times 25 = 500$ surveys. The analysis will be a mean comparison using the independent-samples *t* test on the value of the survey legitimacy scale. Past research has shown that BWCs affect citizen complaints by about 0.29 standard deviations in Florida (calculations available in Appendix A, see Jennings et al., 2015), and a study in Australia found that an intervention which involved altering police conduct affected citizen responses about police legitimacy by about 0.35 standard deviations (calculations available in Appendix A, see Mazerolle et al., 2013). Thus, we assume a Cohen's *d* (1988) effect size of about $\delta = 0.3$. Using the "difference between two independent means" procedure in the G*Power software package (Faul, Erdfelder, Lang, & Buchner, 2007), we calculated the achieved power for our sample and effect size to be 0.92. This result is sensitive to effect size, for example, the power to detect 0.25 standard deviations is about 0.80. The result is more sensitive to survey responses collected, for example, if 10 surveys are collected per officer the power to detect 0.3 standard deviations is about 0.56.

Minimum Detectable Effect Size Given Power and Sample Size

A possible section outlining the MDES may be the following one:

We plan on utilizing 20 officers from the City Police Department and randomizing $n = 10$ officers to wear a BWC (treatment) and $n = 10$ officers to not wear a BWC (control). The officers will hand out about 100 surveys to citizens during the study period, and we expect to receive approximately 25 percent of them for a total sample of $2 \times 10 \times 25 = 500$ surveys. The analysis will be a mean comparison using the independent-samples *t* test on the value of the survey legitimacy scale. Given a desired power level of 0.8, we calculated the minimum detectable effect size in Cohen's d units (1988). Using the "difference between two independent means" procedure in the G*Power software package (Faul et al., 2007), we calculated the minimum detectable effect size for our sample and desired power (0.8) to be 0.25. Past research has shown that BWCs affect citizen complaints by about 0.29 standard deviations in Florida (calculations available in Appendix A, see Jennings et al., 2015), and a study in Australia found that an intervention which involved altering police conduct affected citizen responses about police legitimacy by about 0.35 standard deviations (calculations available in Appendix A, see Mazerolle et al., 2013). Therefore, we are confident that our sample is sensitive enough to detect an effect relevant to the literature. This result is sensitive to sample size, for example, if 10 surveys are collected per officer the minimum detectable effect size is larger, 0.40.

Necessary Size Given Power and Effect Size

A possible section outlining the necessary sample size may be the following one (noting that we are seeking to find the total number of surveys each of the 20 officers needs to hand out):

> We plan on utilizing 20 officers from the City Police Department and randomizing $n = 10$ officers to wear a BWC (treatment) and $n = 10$ officers to not wear a BWC (control). The analysis will be a mean comparison using the independent-samples t test on the value of the survey legitimacy scale. Given a desired power level of 0.8, we calculated the Cohen's d effect size units (1988) that we can expect to be 0.3. Past research has shown that BWCs affect citizen complaints by about 0.29 standard deviations in Florida (calculations available in Appendix A, see Jennings et al., 2015), and a study in Australia found that an intervention which involved altering police conduct affected citizen responses about police legitimacy by about 0.35 standard deviations (calculations available in Appendix A, see Mazerolle et al., 2013). Using the "difference between two independent means" procedure in the G*Power software package (Faul et al., 2007), we calculated the necessary sample size to be 176 in each treatment group. Therefore we need each officer to generate about 18 surveys, on average. This result is sensitive to effect size, for example, if the effect size is 0.2, then each officer needs to collect about 40 surveys.

Cluster Randomized Sample Write-up for Power Given an Effect Size and Sample Design

For brevity, I only present a power analysis that determines power for a given sample design and effect size in this section. For this example, suppose that surveys are distributed to citizens who encounter 125 officers either wearing (treatment) or not wearing (control) a BWC. For each officer, the research team has determined that about 25 surveys will be mailed back during the study period. The surveys are mailed anonymously to the research team and entered into a database. A multilevel mixed model, comparing the surveys nested in officers who are grouped either as treatment or control, will be performed. Again, for simplicity, we assume that the sample will be balanced.

Suppose pilot data were collected to estimate the degree to which survey responses of citizens correlate within officers, and the result was an intraclass correlation of $\rho_{intra} = 0.7$. The pilot analysis also revealed that covariates such as $q = 15$ dummy indicators of the circumstances of the contact at the citizen survey level (e.g., traffic stop) and at the officer level (e.g., proportion of contacts that were traffic stops) explained a quarter of the variance at the citizen level ($R^2_{unit} = 0.25$) and a tenth of the variation at the officer level ($R^2_{cluster} = 0.1$). Given these parameters, a power analysis found that if we assign $n = 125$ officers[1] to treatment and to control that power will exceed 0.8.

A possible section outlining the achieved power may be this one:

We plan on utilizing 250 officers from the City Police Department and randomizing $n = 125$ officers to wear a BWC (treatment) and $n = 125$ officers to not wear a BWC (control). The officers will hand out about 100 surveys to citizens during the study period, and we expect to receive approximately 25 percent of them for a total sample of $2 \times 125 \times 25 = 6,250$ surveys. The analysis will be for a cluster randomized trial with surveys nested within officers. The analysis will be performed using a mixed regression model to estimate the effect of the officer group (treatment vs. control) on the value of the survey legitimacy scale. Past research has shown that BWCs impact citizen complaints by about 0.29 standard deviations in Florida (calculations available in Appendix A, see Jennings et al., 2015), and a study in Australia found that an intervention which involved altering police conduct affected citizen responses about police legitimacy by about 0.35 standard deviations (calculations available in Appendix A, see Mazerolle et al., 2013). Thus, we assume a Cohen's d (1988) effect size of about $\delta = 0.3$.

Pilot data indicate that the intraclass correlation of surveys within officers is about 0.7, but including 30 covariates (15 at the citizen survey level and 15 at the officer level) related to the circumstances of the citizen contact explained 25% of the survey level variance and 10% of the officer level variance. Using these parameters with the formula for a hierarchical design (see page 21, expression 9, in Hedges & Rhoads, 2010) and the RDPOWER software for

---

[1] This must be a large city.

Stata (Hedberg, 2012), we found that this design achieved power of 0.84 for a two-tailed test with $\alpha = 0.05$.

This result is sensitive to effect size, for example, the power to detect 0.25 standard deviations is about 0.69. The result is less sensitive to survey responses collected, for example, if 10 surveys are collected per officer the power to detect 0.3 standard deviations is about 0.83.

Summary

In this chapter, the elements of a power analysis write-up were suggested. We returned to the study example in Chapter 8 about body-worn cameras to write a hypothetical paragraph about a power analysis. Again, the enumerated elements and sample language are only a guide, as different funding or review applications may have differing requirements. It is always a good idea to follow the application's instructions closely.

CHAPTER 10. CONCLUSIONS, FURTHER READING, AND REGRESSION

As an introduction to power analysis, the primary focus of this volume was to outline the interplay between sample design (e.g., sample size) and design parameters (e.g., effect size). A power analysis combines these elements to produce an estimate of the chance of detecting the effect, should it exist. A power analysis can also assume a chance of detecting an effect, and seek to estimate the necessary sample size or minimum detectable effect. The task for this volume was to take one of the more common analyses, the difference in means between two groups, and use it as a case study to explore the dynamics of power analysis.

This chapter attempts to put everything presented in this volume in context. I offer the reason why the focus is on two groups and randomization, and then I offer some leads to other topics in power analysis. Finally, for one last exercise, I discuss power for observational regression designs.

The Case Study of Comparing Two Groups

To accomplish an introduction to power analysis, this volume examined the two-group comparison test at different levels of complexity, from a simple random sample to more complex clustered designs. Both designs considered the use of covariates in a regression framework. For simple random samples without covariates, we also considered the issue of balance (whether the two groups have the same number of observations).

The goal of this exercise was not to give an all-encompassing overview of every kind of analysis that would need an estimate of power. Moreover, this volume did not explore every possible aspect for even the simple two-group case. As is admitted in the footnotes, this book does not cover the implications of stratified samples, sampling weights, or other techniques to examine treatment effects such as matching or propensity scores, and much, much more.

However, the goal of this volume is to introduce the reader to key elements of any power analysis: the need to understand the test statistic and its sampling variance, the effect of controls, and the concept of design effects for deviations from simple random samples. If the reader wishes to use a stratified cluster sample with weights, volumes such Wolter

(2007) should be able to point the way to understanding how different designs' sampling variances create the design effects, and these design effects can be applied to the simple random sample case presented in this volume. Other factors that are important include issues of attrition and/or nonresponse, as these influence the amount of data to be analyzed (unless, of course, the research team employs imputation, which also comes with a design effect).

Further Reading

I have attempted to intersperse this volume with the key citations in this area. In addition to what has been reviewed so far, I offer the following suggestions. There is a lot of work currently under way for understanding power, sample size, and design effects for both randomized and observational designs. The variance of treatment effects across sites is an area that continues to evolve (Rhoads, 2017), and the implications for cluster designs on regression models are now better understood (Lohr, 2014). A classic work in the area of sampling and sample size is Kish (1965), and a more updated volume is Ryan (2013). The primary text on this subject is the seminal work of Cohen (1988). While this volume was critical of the "shirt sizes" approach to effect size expectations, it should be noted that Cohen himself felt responsible (and was critical of) their widespread use (see, e.g., 1994).

This volume attacked most problems with regression formulas, and moved to the ANOVA framework for more complex designs. The regression framework lends itself well to understanding the parameters for power. Readers are encouraged to seek out older social statistics textbooks that do not use matrix algebra in order to understand how to build multivariate regressions from the bivariate relationships (a personal favorite is Blalock, 1972).

Such resources can put other designs into perspective. For example, regression discontinuity (Bloom, 2012) is a popular alternative to randomization because it employs a continuous criteria variable to assign treatment and then fits a multivariate model that includes the criteria and assignment variable to estimate a treatment effect. This treatment effect is essentially a matching estimator, comparing those right before the cutoff with those right after the cutoff. However, power for such designs is complicated precisely because the criteria variable will be correlated with the treatment assignment variable (Schochet, 2009). Working through the bivariate relationships in typical regression will help the reader understand the implications for power when using designs such as regression discontinuity.

Nonlinear Models

This introduction focused on linear outcomes. However, not all dependent variables in social science are linear. Some are binomial (e.g., choices to do something or not), others are count variables (e.g., number of children), and so on. Unlike the normal distribution, models with nonlinear outcomes must be fit with generalized linear models (McCullagh & Nelder, 1989). However, unlike normally distributed variables, the variance of a nonlinear variable is a function of the variable's mean (whereas with normal variables the variance and the mean are unrelated).

The connection between the mean and variance posses problems for power analysis because effects of variables influence means, and in turn influence the variances. There has been a lot of work to solve the problem of power and sample size, in terms of both single-level models (see, e.g., Hsieh, Bloch, & Larsen, 1998) and multilevel models (Moerbeek, Van Breukelen, & Berger, 2001).

More Complex Models

Of course, there are more models and problems in research than simply comparing two groups. Power and sample size estimations have been worked out for structural equation models (see MacCallum, Browne, & Sugawara, 1996), survival models (see, e.g., Lachin & Foulkes, 1986), mediation models (Fritz & MacKinnon, 2007), and many more. A good strategy when searching for such resources is to specify both "power" and "sample size" as different disciplines approach the program from different angles. Another piece of advice is to look outside your specific discipline and to think creatively about any research problem. Power for a recidivism study may be informed by the medical literature's work on power for testing a drug (both can be longitudinal discrete outcomes).

Issues of Cost

Data collection costs money. I often find that in my own research involving primary data collection, a very large portion of the funding goes just to collecting the data, and a very small portion goes to analysis. This makes a good power analysis very important, as it drives most of the resource need. In randomized trials, both single level and multilevel, it is important to understand the cost structure of data collection.

Many times, the cost of a treatment unit will be more expensive than for a control unit (because you do more with the treatment units such as offer a curriculum). For single-level models simply knowing the cost of treatment and control units, along with the effects of not using a balanced design, can be helpful in determining allocation of units to conditions.

For multilevel models, the work of Raudenbush and colleagues has influenced designs in education research (Raudenbush, 1997; Raudenbush & Liu, 2000). Issues to consider are not only the cost of treatment and control units but also the cost of recruiting the clusters themselves, which will differ based on treatment and control conditions (i.e., more work may be needed to keep the control schools). Programs such as Optimal Design (Spybrook, Raudenbush, Liu, Congdon, & Martínez, 2006) employ the same design parameters used for power calculation to help find the optimal mix of treatment and control units and clusters.

Observational Regression

As a final dive into power analysis, I offer a brief look at power analysis for observational studies. Randomized designs were selected for this volume because they (in expectation) remove all the collinearity between the variable of interest and controls, allowing easier computations for the power to detect effects. As we saw in Chapter 5, wiping out such correlations made the calculations and assumptions easier. As you will see below, power analysis for observational studies can be more difficult because you almost have to know the correlation structure of the data before you collect the data.

This does not preclude researchers from power analysis for observational studies, however. For a simple example, consider linear regression. Assuming the model is correctly specified, observational data from a simple random sample can be used with ordinary least squares (OLS) routines to fit either bivariate or multivariate regression models. In fact, our tests in Chapter 4 were simply bivariate regression models with a dichotomous predictor, and the test in Chapter 5 was a multivariate regression with a dichotomous predictor and a continuous control. Both of those power analyses focused on the slope of the dichotomous predictor.

However, power can be computed for *any* of the slopes, the entire model, or even a specific set of slopes. These power computations are more straightforward using the F distribution, so you may wish to review

Chapter 2.[1] To keep the discussion brief, we will consider only scale-free parameters such as correlations and multiple correlations. Interested readers are encouraged to read more advanced volumes such as Liu (2013).

Bivariate Regression

In the bivariate regression model, the test of the regression slope is the same as the test of the correlation coefficient that measures the correlation between two variables (e.g., y and x, noted as ρ). Thus, we can use the F test for the correlation coefficient

$$F = \frac{\rho^2 (N-2)}{1 - \rho^2}.$$ (Equation 10.1)

This F test has 1 degree of freedom in the numerator and $N-2$ degrees of freedom in the denominator. For example, consider the following values:

y	x
4	2
6	9
7	8
5	6

We can compute that $N = 4$ and the correlation is about $\rho_{yx} = 0.876$. The associated F test for this correlation is about

$$F = \frac{0.876^2 (4-2)}{1 - .0.876^2} = 6.598.$$

The p value associated with this test (which has 1 degree of freedom in the numerator and $N - 2 = 2$ in the denominator) is about .124. We can compute another effect size that can be used as the noncentrality parameter for the F distribution. This effect size is f^2 (Cohen, 1988) and is defined for this test as

$$f^2 = \frac{\rho^2}{1 - \rho^2},$$ (Equation 10.2)

---

[1] In fact, a great many tests can be used with the F distribution (Murphy, Myors, & Wolach, 2009).

and the noncentrality parameter for this case is the product of the effect size and total sample size

$$\lambda = f^2 N. \qquad \text{(Equation 10.3)}$$

We can use this formula for the noncentrality parameter (λ) in the noncentral F distribution. Noncentral F distributions work much as noncentral t distributions (except they only have one tail). Like the H function we used for the t test (Equation 4.23), we can use a similar function for the F distribution to find power

$$\beta = G\left[F_{(df_1, df_2)1-\alpha}, df_1, df_2, \lambda\right], \qquad \text{(Equation 10.4)}$$

where $G[a, b, c, d]$ is the cumulative distribution function of the *noncentral* F distribution at point a with b degrees of freedom in the numerator, c degrees of freedom in the denominator, and noncentrality parameter d. Point a is the critical value of the F distribution with df_1 degrees of freedom in the numerator and df_2 in the denominator. In the example above, the critical value of F for $\alpha = 0.05$ where $df_1 = 1$ and $df_2 = 2$ is about 18.513. Note that this is the square of $t_{(2)0.975} = 4.303$ in the Appendix.

Using the effect size of $f^2 = \frac{\rho^2}{1-\rho^2} = \frac{.876^2}{1-.876^2} = 3.299$ and the sample size $N = 4$ to produce $\lambda = 3.299 \times N = 13.196$ as the noncentrality parameter and a computer, we can find that the Type II error for this test is about 0.499 and the power is 0.501. This procedure can be used to test power for the slope, the correlation, and the whole model (it is all the same result).

The Entire Regression Model

An entire regression model is summarized by the R^2 multiple correlation coefficient, which reports the proportion of the variance in the outcome that is explained by the set of covariates as specified in the model. Given a set number of q predictors, the statistical test of the multiple correlation coefficient R^2 is very similar in form to Equation 10.1:

$$F = \frac{R^2 q}{\left(1 - R^2\right)\left(N - q - 1\right)}. \qquad \text{(Equation 10.5)}$$

This test employs q degrees of freedom in the numerator and $N - q - 1$ degrees of freedom in the denominator. Thus, if researchers have a sense as to what R^2 will be in their data, they can estimate the power of their design. We use the degrees of freedom, as before, to obtain the critical value of the F distribution. Next, the effect size is essentially the same as for the single predictor case:

$$f^2 = \frac{R^2}{1 - R^2}. \qquad \text{(Equation 10.6)}$$

For example, suppose we believe that for a model with 3 predictors and 15 observations the R^2 will be 0.25. The critical F value for a test with $\alpha = 0.05$ is $F_{3,11} = 3.587$. The effect size is

$$f^2 = \frac{0.25}{1 - 0.25} = .333,$$

which leads to a noncentrality parameter of $\lambda = 0.333 \times 15 = 4.995$, which is associated with a Type II error of 0.679 and power of 0.321.

A Block of Predictors

Suppose a research team has a good idea that the set of c controls will capture a portion of the variation, which we note as $R^2_{control}$. However, this research team wants to test whether additional k variables of interest will (along with the controls) explain a larger portion of the variation, which we note as R^2_{total}. The test of this new block of variables is also an F test:

$$F = \frac{\left(R^2_{total} - R^2_{control}\right) / k}{\left(1 - R^2_{total}\right) / (N - k - c - 1)}.$$

This test employs k degrees of freedom in the numerator and $N - k - c - 1$ degrees of freedom in the denominator. Thus, if researchers have a sense as to what R^2_{total} and what $R^2_{control}$ will be in their data, they can estimate the power of their design. The effect size for this design is

$$f^2 = \frac{\left(R^2_{total} - R^2_{control}\right)}{1 - R^2_{total}}. \qquad \text{(Equation 10.7)}$$

For example, suppose that researchers have 40 observations, and that the full model will explain 30% of the variation ($R^2_{total} = 0.3$) but the model with only $c = 4$ controls will explain about 20% of the variation ($R^2_{control} = 0.2$). What is the power for a model with the additional $k = 2$ predictors? The critical F value for a test with $\alpha = 0.05$ is $F_{2,33} = 3.285$. The effect size is

$$f^2 = \frac{0.3 - 0.2}{1 - 0.3} = 0.143,$$

which leads to a noncentrality parameter of $\lambda = 0.143 \times 40 = 5.720$, which is associated with a Type II error of 0.478 and power of 0.522.

122

Summary

This chapter offers some context to the volume. The reader was informed as to the purpose of the book and why two-group studies were the focus. The chapter also includes more citations to literature about other types of analyses that may require a sound power analysis. Finally, the volume ends with a final dive into power analyses, this time for observational regression.

APPENDIX

Table 1   Body Mass Index (BMI) Data

	TREAT	BMI	PRE_BMI	TREAT	BMI	PRE_BMI
1	0	37.66	38.08	1	28.62	28.84
2	0	32.61	31.43	1	35.91	36.10
3	0	23.34	25.38	1	27.25	27.81
4	0	43.75	43.27	1	30.06	31.98
5	0	37.25	35.76	1	36.26	36.34
6	0	32.36	32.53	1	28.01	27.15
7	0	36.35	36.81	1	28.69	29.45
8	0	25.21	26.17	1	27.34	29.90
9	0	30.21	29.24	1	26.87	29.70
10	0	32.82	34.03	1	26.81	26.61
11	0	39.40	39.98	1	39.79	39.59
12	0	38.75	38.26	1	37.87	37.95
13	0	37.43	38.65	1	41.20	40.59
14	0	35.56	34.73	1	26.22	26.09
15	0	29.40	31.17	1	27.71	27.01
16	0	28.62	27.55	1	27.05	26.52
17	0	40.75	41.43	1	35.20	35.08
18	0	38.58	38.95	1	27.64	26.56
19	0	43.37	42.34	1	34.29	34.07
20	0	33.87	33.84	1	27.53	28.58
21	0	37.42	37.53	1	29.15	28.58
22	0	28.08	28.39	1	26.27	25.93
23	0	42.41	42.16	1	43.97	45.54
24	0	31.33	31.29	1	38.65	37.38
25	0	46.98	45.03	1	28.12	27.30

Table 2 Math Data

	SCHOOL	TREAT	MATH	PRE_MATH	MEAN_PRE_MATH
1	113	1	7	1	2.25
2	113	1	6	1	2.25
3	113	1	6	3	2.25
4	113	1	7	4	2.25
5	116	1	11	4	4.00
6	116	1	10	4	4.00
7	116	1	8	3	4.00
8	116	1	11	5	4.00
9	117	0	3	4	4.75
10	117	0	9	5	4.75
11	117	0	6	5	4.75
12	117	0	8	5	4.75
13	123	1	9	5	4.50
14	123	1	11	4	4.50
15	123	1	8	4	4.50
16	123	1	10	5	4.50
17	210	1	9	3	3.50
18	210	1	6	3	3.50
19	210	1	7	5	3.50
20	210	1	8	3	3.50
21	226	0	9	5	3.50
22	226	0	2	1	3.50
23	226	0	10	4	3.50
24	226	0	6	4	3.50
25	303	0	6	3	3.00
26	303	0	5	3	3.00
27	303	0	7	4	3.00
28	303	0	3	2	3.00
29	304	1	7	3	3.25
30	304	1	8	1	3.25
31	304	1	10	4	3.25
32	304	1	10	5	3.25
33	319	0	3	1	3.00
34	319	0	10	4	3.00
35	319	0	7	4	3.00
36	319	0	8	3	3.00
37	321	0	4	4	3.25
38	321	0	3	4	3.25
39	321	0	5	2	3.25
40	321	0	8	3	3.25

Table 3 Simulated Multisite Randomized Trial Data

	K	Y	TREAT	X	MEAN_X
1	1	49.85	0	44.45	45.87
2	1	68.68	1	40.25	45.87
3	1	93.58	1	71.46	45.87
4	1	94.87	1	55.80	45.87
5	1	53.35	0	34.70	45.87
6	1	76.06	1	56.06	45.87
7	1	38.84	0	34.58	45.87
8	1	52.51	0	29.65	45.87
9	2	56.37	0	50.83	45.59
10	2	41.12	1	44.59	45.59
11	2	43.87	0	41.55	45.59
12	2	61.60	0	71.08	45.59
13	2	64.61	1	34.08	45.59
14	2	41.16	1	44.80	45.59
15	2	49.19	0	38.13	45.59
16	2	53.13	1	39.69	45.59
17	3	51.04	0	40.64	50.65
18	3	70.55	0	51.20	50.65
19	3	66.62	1	40.00	50.65
20	3	63.24	1	64.75	50.65
21	3	52.30	0	48.79	50.65
22	3	57.85	1	58.03	50.65
23	3	98.13	1	68.91	50.65
24	3	63.27	0	32.84	50.65
25	4	68.89	1	63.37	58.01
26	4	74.77	0	77.26	58.01
27	4	61.42	1	32.17	58.01
28	4	84.67	1	45.24	58.01
29	4	87.34	0	70.82	58.01
30	4	59.89	1	52.67	58.01
31	4	70.72	0	64.13	58.01
32	4	71.49	0	58.42	58.01
33	5	41.27	0	47.87	49.88
34	5	68.86	1	63.58	49.88
35	5	49.65	1	23.55	49.88
36	5	23.52	0	54.59	49.88
37	5	50.51	0	65.20	49.88
38	5	50.07	0	43.27	49.88
39	5	46.49	1	37.74	49.88
40	5	80.35	1	63.26	49.88

Table 4 Quantiles of t Distribution by Degrees of Freedom (df) and Standard Normal Distribution ($df = \infty$)

df	Quantile						
	$t_{(df)0.1}$	$t_{(df)0.2}$	$t_{(df)0.3}$	$t_{(df)0.95}$	$t_{(df)0.975}$	$t_{(df)0.99}$	$t_{(df)0.995}$
2	−1.886	−1.061	−0.617	2.92	4.303	6.965	9.925
3	−1.638	−0.978	−0.584	2.353	3.182	4.541	5.841
4	−1.533	−0.941	−0.569	2.132	2.776	3.747	4.604
5	−1.476	−0.92	−0.559	2.015	2.571	3.365	4.032
6	−1.44	−0.906	−0.553	1.943	2.447	3.143	3.707
7	−1.415	−0.896	−0.549	1.895	2.365	2.998	3.499
8	−1.397	−0.889	−0.546	1.86	2.306	2.896	3.355
9	−1.383	−0.883	−0.543	1.833	2.262	2.821	3.25
10	−1.372	−0.879	−0.542	1.812	2.228	2.764	3.169
11	−1.363	−0.876	−0.54	1.796	2.201	2.718	3.106
12	−1.356	−0.873	−0.539	1.782	2.179	2.681	3.055
13	−1.35	−0.87	−0.538	1.771	2.16	2.65	3.012
14	−1.345	−0.868	−0.537	1.761	2.145	2.624	2.977
15	−1.341	−0.866	−0.536	1.753	2.131	2.602	2.947
16	−1.337	−0.865	−0.535	1.746	2.12	2.583	2.921
17	−1.333	−0.863	−0.534	1.74	2.11	2.567	2.898
18	−1.33	−0.862	−0.534	1.734	2.101	2.552	2.878
19	−1.328	−0.861	−0.533	1.729	2.093	2.539	2.861
20	−1.325	−0.86	−0.533	1.725	2.086	2.528	2.845
21	−1.323	−0.859	−0.532	1.721	2.08	2.518	2.831
22	−1.321	−0.858	−0.532	1.717	2.074	2.508	2.819
23	−1.319	−0.858	−0.532	1.714	2.069	2.5	2.807
24	−1.318	−0.857	−0.531	1.711	2.064	2.492	2.797
25	−1.316	−0.856	−0.531	1.708	2.06	2.485	2.787
30	−1.31	−0.854	−0.53	1.697	2.042	2.457	2.75
35	−1.306	−0.852	−0.529	1.69	2.03	2.438	2.724
40	−1.303	−0.851	−0.529	1.684	2.021	2.423	2.704
45	−1.301	−0.85	−0.528	1.679	2.014	2.412	2.69
50	−1.299	−0.849	−0.528	1.676	2.009	2.403	2.678
55	−1.297	−0.848	−0.527	1.673	2.004	2.396	2.668
60	−1.296	−0.848	−0.527	1.671	2	2.39	2.66
65	−1.295	−0.847	−0.527	1.669	1.997	2.385	2.654
70	−1.294	−0.847	−0.527	1.667	1.994	2.381	2.648
75	−1.293	−0.846	−0.527	1.665	1.992	2.377	2.643
80	−1.292	−0.846	−0.526	1.664	1.99	2.374	2.639
85	−1.292	−0.846	−0.526	1.663	1.988	2.371	2.635
90	−1.291	−0.846	−0.526	1.662	1.987	2.368	2.632
95	−1.291	−0.845	−0.526	1.661	1.985	2.366	2.629
100	−1.29	−0.845	−0.526	1.66	1.984	2.364	2.626
	$z_{0.1}$	$z_{0.2}$	$z_{0.3}$	$z_{0.95}$	$z_{0.975}$	$z_{0.99}$	$z_{0.995}$
∞	−1.282	−0.842	−0.524	1.645	1.96	2.326	2.576

REFERENCES

Blalock, H. M. (1972). *Social statistics* (2nd ed.). New York, NY: McGraw-Hill.

Bloom, H. S. (1995). Minimum detectable effects a simple way to report the statistical power of experimental designs. *Evaluation Review, 19*, 547–556.

Bloom, H. S. (2005). *Learning more from social experiments: Evolving analytic approaches.* New York, NY: Russell Sage Foundation.

Bloom, H. S. (2012). Modern regression discontinuity analysis. *Journal of Research on Educational Effectiveness, 5*, 43–82.

Borenstein, M., Hedges, L. V., Higgins, J., & Rothstein, H. R. (2009). *Introduction to meta-analysis.* Chichester, England: Wiley.

Brown, S. R., & Melamed, L. E. (1990). *Experimental design and analysis.* Thousand Oaks, CA: Sage.

Casella, G. (2008). *Statistical design.* New York, NY: Springer Science+Business Media.

Cohen, J. (1988). *Statistical power analysis for the behavioral sciences.* London, England: Routledge.

Cohen, J. (1992). A power primer. *Psychological Bulletin, 112*, 155–159.

Cohen, J. (1994). The earth is round ($p < .05$). *American Psychologist, 49*, 997–1003.

Dahl, D. B. (2009). *xtable: Export tables to LaTeX or HTML (R package version).*

Dhurandhar, E. J., Dawson, J., Alcorn, A., Larsen, L. H., Thomas, E. A., Cardel, M., … Allison, D. B. (2014). The effectiveness of breakfast recommendations on weight loss: A randomized controlled trial. *American Journal of Clinical Nutrition, 100*, 507–513.

Donner, A., & Klar, N. (2000). *Design and analysis of cluster randomization trials in health research.* London, England: Arnold.

Easterbrook, P. J., Gopalan, R., Berlin, J., & Matthews, D. R. (1991). Publication bias in clinical research. *Lancet, 337*, 867–872.

Faul, F., Erdfelder, E., Lang, A.-G., & Buchner, A. (2007). G* power 3: A flexible statistical power analysis program for the social, behavioral, and biomedical sciences. *Behavior Research Methods, 39*, 175–191.

Fox, J. (2015). *Applied regression analysis and generalized linear models.* Thousand Oaks, CA: Sage.

Fritz, M. S., & MacKinnon, D. P. (2007). Required sample size to detect the mediated effect. *Psychological Science, 18*, 233–239.

Gamoran, A., Turley, R. N. L., Turner, A., & Fish, R. (2012). Differences between Hispanic and non-Hispanic families in social capital and child development: First-year findings from an experimental study. *Research in Social Stratification and Mobility, 30*, 97–112.

Gulliford, M. C., Ukoumunne, O. C., & Chinn, S. (1999). Components of variance and intraclass correlations for the design of community-based surveys and intervention studies: Data from the Health Survey for England 1994. *American Journal of Epidemiology, 149*, 876–883.

Hedberg, E. C. (2012). *RDPOWER: Stata module to perform power calculations for random designs (Statistical Software Components).* Chestnut Hill, MA: Boston College Department of Economics.

Hedberg, E. C., & Hedges, L. V. (2014). Reference values of within-district intraclass correlations of academic achievement by district characteristics: Results from a meta-analysis of district-specific values. *Evaluation Review, 38*, 546–582.

128

Hedges, L. V. (2007). Effect sizes in cluster-randomized designs. *Journal of Educational and Behavioral Statistics, 32,* 341–370.

Hedges, L. V., & Hedberg, E. C. (2007). Intraclass correlation values for planning group-randomized trials in education. *Educational Evaluation and Policy Analysis, 29,* 60–87.

Hedges, L. V., & Hedberg, E. C. (2013). Intraclass correlations and covariate outcome correlations for planning two- and three-level cluster-randomized experiments in education. *Evaluation Review, 37,* 445–489.

Hedges, L. V., & Rhoads, C. (2010). *Statistical power analysis in education research.* Washington, DC: National Center for Special Education Research.

Hill, C. J., Bloom, H. S., Black, A. R., & Lipsey, M. W. (2008). Empirical benchmarks for interpreting effect sizes in research. *Child Development Perspectives, 2,* 172–177.

Hsieh, F. Y., Bloch, D. A., & Larsen, M. D. (1998). A simple method of sample size calculation for linear and logistic regression. *Statistics in Medicine, 17,* 1623–1634.

Ioannidis, J. P. A. (2005). Why most published research findings are false. *PLoS Medicine, 2,* e124.

Jennings, W. G., Lynch, M. D., & Fridell, L. A. (2015). Evaluating the impact of police officer body-worn cameras (BWCs) on response-to-resistance and serious external complaints: Evidence from the Orlando Police Department (OPD) experience utilizing a randomized controlled experiment. *Journal of Criminal Justice, 43,* 480–486.

Kirk, R. E. (1995). *Experimental design* (3rd ed.). Pacific Grove, CA: Brooks/Cole.

Kish, L. (1965). *Survey sampling.* New York, NY: Wiley.

Lachin, J. M., & Foulkes, M. A. (1986). Evaluation of sample size and power for analyses of survival with allowance for nonuniform patient entry, losses to follow-up, noncompliance, and stratification. *Biometrics, 42,* 507–519.

Leifeld, P. (2013). texreg: Conversion of statistical model output in R to LaTeX and HTML tables. *Journal of Statistical Software, 55*(8), 1–24.

Lipsey, M. W. (1990). *Design sensitivity: Statistical power for experimental research.* Newbury Park, CA: Sage.

Liu, X. S. (2013). *Statistical power analysis for the social and behavioral sciences: Basic and advanced techniques.* London, England: Routledge.

Lohr, S. L. (2009). *Sampling: Design and analysis.* Boston, MA: Cengage Learning.

Lohr, S. L. (2014). Design effects for a regression slope in a cluster sample. *Journal of Survey Statistics and Methodology, 2,* 97–125.

MacCallum, R. C., Browne, M. W., & Sugawara, H. M. (1996). Power analysis and determination of sample size for covariance structure modeling. *Psychological Methods, 1,* 130–149.

Mazerolle, L., Antrobus, E., Bennett, S., & Tyler, T. R. (2013). Shaping citizen perceptions of police legitimacy: A randomized field trial of procedural justice. *Criminology, 51,* 33–63.

McCullagh, P., & Nelder, J. A. (1989). *Generalized linear models* (2nd ed.). London, England: Chapman & Hall.

McCulloch, C. E., & Searle, S. R. (2001). *Generalized, linear, and mixed models.* Hoboken, NJ: Wiley.

Moerbeek, M., Van Breukelen, G. J. P., & Berger, M. P. F. (2001). Optimal experimental designs for multilevel logistic models. *Journal of the Royal Statistical Society: Series D (The Statistician), 50,* 17–30.

Murphy, K., Myors, B., & Wolach, A. (2009). *Statistical power analysis: A simple and general model for traditional and modern hypothesis tests.* London, England: Routledge.

129

Murray, D. M. (1998). *Design and analysis of group-randomized trials.* New York, NY: Oxford University Press.

Murray, D. M., Rooney, B. L., Hannan, P. J., Peterson, A. V., Ary, D. V., Biglan, A., ... Schinke, S. P. (1994). Intraclass correlation among common measures of adolescent smoking: Estimates, correlates, and applications in smoking prevention studies. *American Journal of Epidemiology, 140,* 1038–1050.

Neyman, J., & Pearson, E. S. (1933). The testing of statistical hypotheses in relation to probabilities a priori. *Mathematical Proceedings of the Cambridge Philosophical Society, 29,* 492–510.

O'Connell, A. A., & McCoach, D. B. (Eds.). (2008). *Multilevel modeling of educational data.* Charlotte, NC: Information Age.

Open Science Collaboration. (2015). Estimating the reproducibility of psychological science. *Science, 349,* aac4716.

Porter, A. C., & Raudenbush, S. W. (1987). Analysis of covariance: Its model and use in psychological research. *Journal of Counseling Psychology, 34,* 383–392.

Raudenbush, S. W. (1997). Statistical analysis and optimal design for cluster randomized trials. *Psychological Methods, 2,* 173–185.

Raudenbush, S. W., & Bryk, A. S. (2002). *Hierarchical linear models: Applications and data analysis methods.* Thousand Oaks, CA: Sage.

Raudenbush, S. W., Bryk, A. S., & Congdon, R. (2004). *HLM 6 for Windows [Computer software].* Lincolnwood, IL: Scientific Software International.

Raudenbush, S. W., & Liu, X.-F. (2000). Statistical power and optimal design for multisite randomized trials. *Psychological Methods, 5,* 199–213.

Rhoads, C. (2017). Coherent power analysis in multilevel studies using parameters from surveys. *Journal of Educational and Behavioral Statistics, 42,* 166–194.

Rice, J. (2006). *Mathematical statistics and data analysis.* Boston, MA: Cengage Learning.

Ryan, T. P. (2013). *Sample size determination and power.* Hoboken, NJ: Wiley.

Schochet, P. Z. (2008). Statistical power for random assignment evaluations of education programs. *Journal of Educational and Behavioral Statistics, 33,* 62–87.

Schochet, P. Z. (2009). Statistical power for regression discontinuity designs in education evaluations. *Journal of Educational and Behavioral Statistics, 34,* 238–266.

Schultz, E. (1955). Rules of thumb for determining expectations of mean squares in analysis of variance. *Biometrics, 11,* 123–135.

Sharpsteen, C., & Bracken, C. (2013). *tikzdevice: R graphics output in LaTeX format. R package version 0.7.0.* Retrieved from http://CRAN.R-project.org/package=tikzDevice.

Spybrook, J. (2007). *Examining the experimental designs and statistical power of group randomized trials funded by the Institute of Education Sciences* (Unpublished doctoral dissertation). University of Michigan, Ann Arbor.

Spybrook, J., Raudenbush, S. W., Liu, X.-F., Congdon, R., & Martínez, A. (2006). *Optimal design for longitudinal and multilevel research: Documentation for the "optimal design" software.* Ann Arbor: University of Michigan School of Education, Hierarchical Models Project.

Student. (1908). The probable error of a mean. *Biometrika, 6,* 1–25.

Westine, C. D., Spybrook, J., & Taylor, J. A. (2013). An empirical investigation of variance design parameters for planning cluster-randomized trials of science achievement. *Evaluation Review, 37,* 490–519.

White, M. D. (2014). *Police officer body-worn cameras: Assessing the evidence.* Washington, DC: Office of Community-Oriented Policing Services.

Wildt, A. R., & Ahtola, O. (1978). *Analysis of covariance.* Newbury Park, CA: Sage.

Wolter, K. M. (2007). *Introduction to variance estimation* (2nd ed.). New York, NY: Springer Science+Business Media.

Ziliak, S. T., & McCloskey, D. N. (2008). *The cult of statistical significance: How the standard error costs us jobs, justice, and lives.* Ann Arbor: University of Michigan Press.

INDEX

132

134